THE DELIVERY of URBAN SERVICES

Outcomes of Change

Volume 10, URBAN AFFAIRS ANNUAL REVIEWS

THE DELIVERY of URBAN SERVICES

Outcomes of Change

Edited by
ELINOR OSTROM

Volume 10, URBAN AFFAIRS ANNUAL REVIEWS

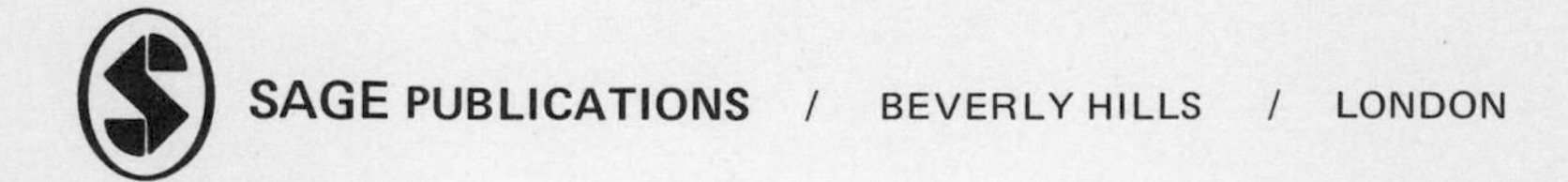

For information address:

SAGE PUBLICATIONS, INC.
275 South Beverly Drive
Beverly Hills, California 90212

SAGE PUBLICATIONS LTD
St George's House / 44 Hatton Garden
London EC1N 8ER

Printed in the United States of America

International Standard Book Number 0-8039-0469-X (cloth)
0-8039-0679-X (paper)

Library of Congress Catalog Card No. 75-11133

FIRST PRINTING

CONTENTS

Introduction

ELINOR OSTROM

□ CHANGES IN THE INSTITUTIONAL ARRANGEMENTS for delivering urban goods and services create incentives and deterrents. Payoffs to participants may shift. Participants may, as a result, behave in predicted or unpredicted ways. When the resulting behavior is similar to that predicted, the institutional change is likely to be a success. But far too many urban programs are based on inadequate analysis of how institutional arrangements affect the strategic considerations of various participants. Surprising and frequently tragic results have followed. Failure, in many cases, leads to the adoption of another program—one often based, as was the first, on inadequate analysis of the strategic behavior of the different actors. Failure seems to breed failure.

In light of a rather gloomy record in improving urban delivery systems, the authors of this volume have each asked: "What do social scientists know that can be useful to public officials and citizens in organizing the delivery of urban services more effectively?" Each writer tackles this question in a different way and for different delivery systems. None offers a panacea. Nevertheless, two general themes emerge from the "collective wisdom" of this group of scholars.

First, institutional arrangements *do* make a difference. Although all the authors would acknowledge social and economic factors as important influences on the delivery systems for urban public goods, the general structure of the delivery system itself is *also* seen as an essential variable affecting outcomes. This is not startling. But the recent literature of social science has devoted much space to the question of whether political structures have any impact.

The second theme concerns diversity. The wide range of institutional arrangements found in America's metropolitan areas is seen as an asset. Reliance on any single set of decision rules exposes all to the risk of total institutional failure. In an imperfect world where institutions are filled with weaknesses, redundancy in organizational arrangements may prevent the failure of any one set of decision rules from seriously handicapping us, as citizens, in accomplishing some of our goals. Multiplicity of arrangements also enables us to test the relative performance of different types of institutional practices and to thus evolve new solutions to different kinds of problems. These chapters reflect the growing possibility of developing a cumulative, empirical theory of institutional analysis and design.

Public housing is one example. Roger S. Ahlbrandt focuses on the incentives for different actors in the delivery of governmentally assisted housing programs since the 1930s. In the main, he paints a grim picture. Except for some recent programs, the incentives have led to actions that culminate in an inefficient delivery system. Members of the U.S. Congress are motivated to improve their chances for reelection. As a result, they generally support programs mounted by strong lobbying. And since the costs of lobbying are high—particularly for groups such as tenants—the construction industry and allied unions have had considerable influence on the shape of federal legislation. The potential beneficiaries of these programs have had little.

HUD officials, Ahlbrandt argues, have been motivated to resist change. Their complex goals include prestige, professionalism, budget size, and reduction of conflict. These goals have led them to develop a complex set of rules and regulations. Local housing authority officials have had similar incentives, but they have also been more exposed to direct confrontation by well-organized local interests. Investors have been motivated to promote projects having low risks, thus avoiding problem populations and locations. Builders have, of course, seen projects as business opportunities. The cumulative effect has been the development of a system that results in considerably higher per unit costs for federally subsidized housing than for corresponding units produced in the regular market.

Ahlbrandt then describes a new institutional arrangement, Neighborhood Housing Services (NHS), that delivers housing in the context of neighborhood preservation. Its incentives, he observes, tend to produce actions efficient in terms of both production and consumption, or supply and demand. He argues for further study of the incentives developed in NHS, so that the future design of other governmentally assisted housing programs can have a higher probability of efficient results.

Dennis R. Young and Janet Fink examine juvenile corrections. They focus on the ways organization impinges on demand and supply *decisions* and, thus, on outcomes. They see the juvenile court as an institution expressing the demand for juvenile services, and they ask whether the court operates effectively as a "proxy-consumer" of juvenile correctional services.

For a juvenile court to effectively demand this public service, judges must have expert evaluation and information at their disposal so that they may intelligently select among alternate services available. Young and Fink discuss a small pilot study that examined this relationship, and they find that juvenile court judges are dependent on the probation service. This service, however, acts as both the enforcement arm and the information source for the court. Few juvenile judges have much specialized training other than law school to prepare them for these duties. The time available to such judges for any single case is extraordinarily limited due to their heavy case loads. A relatively wide range of alternative facilities, however, is available to most of these judges in their attempt to tailor the treatment of each juvenile to the problems of the case. In light of their analysis, Young and Fink present several ideas for change which would increase the capabilities of juvenile court judges to act more effectively in their role as representatives of the demand site for this public service.

Robert L. Bish reviews school finance reform and fiscal equalization. He discusses the issues raised in several recent court cases and analyzes the likely consequences of district power equalizing. In *Serrano vs. Priest,* the judge accepted district power equalizing as one of four reforms that would satisfy the court. It was proposed as one method of removing the influence of school district wealth on expenditures for educating children in a district. This relatively new institutional device has received much support, since local control of expenditures is still permitted even though the influence of district wealth is supposedly eliminated.

But Bish challenges the assumption that high district wealth results in low tax rates and high expenditures. He challenges a related assumption, that property values are not affected by changes in taxes and expenditures for public services. Drawing upon land rent theory, he demonstrates how power equalization results in one-shot increases or decreases in property values but fails to produce the expected long-range shifts in welfare. He shows, for example, how power equalization can adversely affect poor residents who live in school districts having an industrial property base. In this situation, their tax burden might rise, and the amount spent on education might fall. Bish is not arguing against the need for redistributing the costs of education in a more equitable manner. He is, however, urging more careful analysis before adopting programs that will not accomplish their objectives.

Luvern L. Cunningham, an astute scholar and teacher, takes a different perspective on institutional arrangements. He has been an active participant in an on-going social experiment in third-party problem-solving related to the Detroit Public Schools. He analyzes the role of such groups and discusses what they can and cannot do. By presenting the history of the Detroit Education Task Force, he provides us with a series of important lessons that have considerable applicability to fields beyond those of primary and secondary education. "Most of the energy available in a large bureaucratic enterprise and its governance system," he asserts, "is expended on adult concerns, not the concerns of children and youth." We could generalize that statement and ask ourselves whether most of the energy in a large bureaucratic enterprise focuses on the problems of those who produce the public service rather than those who consume it.

Cunningham's lesson—that large agencies are not very effective in representing the preferences of the clients they serve—is substantiated by Michael P. Smith and Douglas D. Rose, but with a much different service and location. Smith and Rose surveyed citizens to learn about their preferences for and participation in the state of Washington's water policy. They found that the citizens preferred a wide variety of modes of representation, including those of experts and administrators as well as the general public. The authors found that citizen participation cannot be viewed as a means of coopting the public into accepting nonpreferred policy output. Satisfaction with outcomes is related to the closeness of fit between public preferences and policy outcomes.

Robert B. Hawkins, Jr., writing on special districts, reports that California residents also prefer a wide diversity in local forms of decision-making. He observes that residents of smaller communities —the citizens most frequently served by the largest number of overlapping special districts—appear *less* confused about the structure of local government, rather than more confused as many critics have assumed. Hawkins examines other arguments for special districts and the empirical evidence in regard to responsiveness, efficiency, and quality of services.

Donald Zauderer examines the effects of one form of institutional arrangement—that of the ballot form—on the voting behavior of urban voters. Since citizen preferences for urban public goods and services must be translated to a large degree through the ballot, Zauderer asks what differences the form of the ballot makes on the process of making choices. He examines the effects of changing from a single-choice party column ballot to an office type ballot on vote roll-off. Vote roll-off is the decrease in combined party vote for all offices on the ballot. Ohio election data from 1934 to 1964 are examined. Zauderer finds that vote roll-off is greatest in electoral districts with high percentages of less educated and minority voters. Their reduced influence in the election of secondary offices has considerable implications for the relationship between their preferences and urban service delivery.

The closing chapters are from the opening phase of a set of studies funded by the RANN Division of the National Science Foundation. In a program solicitation for "Division-Related Research on the Organization of Service Delivery in Metropolitan Areas," NSF announced its willingness to fund research in individual service areas to "provide the knowledge for improving delivery of selected municipal services by describing, analyzing and evaluating alternative organizational arrangements for service delivery in metropolitan areas. . . ."

Much of the work in the first phase of each of these studies was directed at describing the current arrangements for providing each service. For example, E. S. Savas, principal investigator of the study on solid-waste management, presents the conceptual scheme his research team used in their analysis. Four elements specify the delivery system: the service recipient, the service provider, the service arranger, and the service type. A four-dimensional matrix is thus needed to describe the delivery system within a metropolitan area. This results in a much more complex notion of service delivery than has been typically considered. His chapter summarizes data for more than 2,000 cities.

Philip B. Coulter, Lois MacGillivray, and William Edward Vickery examine the organizational and environmental factors that affect fire effectiveness, costs, and productivity. They consider both prevention effectiveness (number of fires per thousand population) and suppression effectiveness (property loss in dollars per thousand population). For costs they use annual fire department budget per capita. For their measure of productivity they add the costs of property loss to the expenditure levels for a jurisdiction to get total cost per capita. More productive departments have a lower total cost per capita than do less productive departments.

Using discriminant analysis, they examine environmental and organizational factors of 324 cities with populations of over 25,000 persons, cities located in 50 small to medium-sized SMSAs. They find that both environmental and organizational variables are associated with cost, effectiveness, and productivity measures. Their findings include some surprises. For example, they find that municipalities having the highest productivity hire fewer fire employees, are less likely to have a full-time paid fire chief, have less comprehensive inspection programs, and have smaller inspection staffs. But they also found that departments in the highest productivity group also have better environmental conditions, which complicates their analyses.

Roger B. Parks examines the implications of altering the current structure of institutional arrangements for the provision of police services in metropolitan areas as recommended by several national commissions. Utilizing a measure of police activity—patrol density—he finds that smaller agencies generally provide considerably higher patrol densities than do larger agencies. But smaller agencies are also less likely to produce auxiliary police services. Some have thus assumed that residents served by small agencies do not receive auxiliary police services. However, Parks briefly examines the patterns of interactions among police agencies in metropolitan areas and shows that these services are most frequently produced by large and, in many instances, specialized agencies, for all of the police agencies in the metropolitan area. He raises the question of whether the recommended reforms will indeed improve performance. He wonders why specialization in the provision of patrol services and routine investigation services has been rejected as an effective organizational arrangement while specialization in the provision of auxiliary services is recommended.

Patrick O'Donoghue examines the influence of physicians' fee levels, medical education programs, and regional factors on the distribution and concentration of primary care physicians. He finds that the income profile of a metropolitan area and other commonly used socioeconomic and demographic measures are not important predictors of physician concentrations. In regard to policy implications, O'Donoghue argues that a network of generic factors determines physician distribution. His findings support the contention that physicians' fee levels can be tools for producing desired alternatives in physical distribution. In designing future medical insurance programs, the recognition of the importance of fee structures may at least enable policy makers to avoid institutional arrangements which accentuate undesirable physician distribution patterns.

1

Governmentally Assisted Housing: Institutions and Incentives

ROGER S. AHLBRANDT, Jr.

INTRODUCTION

□ GOVERNMENTAL INTERVENTION into the field of housing and community development has not been a resounding success. Controversy has surrounded many of the programs, output has not been significant in terms of need, and the clientele served has often expressed dissatisfaction with the product. Arguments can be made that the level of program funding has been inadequate, thereby explaining away many of the criticisms. While it is true that housing and related programs have accounted for a small percentage of total federal expenditures, less than 2% in 1975, this knowledge does not offer a solution to past or present problems. Budgetary priorities are established through a system which takes into account the expressed preferences of the voters, including well-entrenched lobbying groups, and, given the history of the relative low priority attached to housing and community development, it is unlikely that the relative share of the budget allocated to these programs will increase significantly in the future. However, even if budgetary authority for housing increased, results would not be expected to improve as long as the structure of institutional arrangements and incentives remained unchanged. Therefore, program analysis must concentrate upon the delivery system if performance is to be upgraded.

The purpose of this chapter is to examine alternative institutional arrangements for the provision of governmentally assisted housing. The analysis will focus upon the incentive structure and its effect on the goals and objectives of each of the decision makers in the delivery system. The efficiency of various institutional arrangements will be analyzed from both a demand and a supply standpoint. The relationship between the underlying incentives and the expected output of the delivery system will be generalized to provide a framework to assess the merits of alternative institutional arrangements.

A large part of the chapter is concerned with housing policy as legislated by the federal government. Local government, however, may have a significant influence on the quality of the total living environment of which housing is just a part. Therefore, the last section of the chapter turns to an analysis of housing programming from the perspective of local government and concentrates upon the development of a viable neighborhood preservation strategy in light of the housing assistance provided by the federal government.

As used herein, housing services are more than the services provided by a physical unit. The unit exists in a neighborhood and the quality of life in that location influences the household's valuation of the services provided. Literature has discussed the dimensions of the housing environment in terms of social relationships, the physical structure, security, management, and the neighborhood environment.[1] These considerations need to be taken into account in an evaluation of a given delivery system and its relative efficiency.

Efficiency has both demand and supply aspects and must be analyzed accordingly (Bish and V. Ostrom, 1973, chapter 3; E. Ostrom, 1972). The supply side is concerned with production efficiencies, i.e., the extent to which costs are minimized for a given output level holding quality constant. Demand considerations concern the valuation of the resulting output level by the consumer, the tenant, or the owner of a subsidized unit.

The demand side is particularly relevant to those goods which are provided through a quasi-nonmarket decision-making structure. As production decisions are removed from the constraints of the marketplace, the services provided may reflect the preferences or biases of the producer rather than of the consumer (Bish and Warren, 1972). This situation may occur in the provision of federally assisted housing where the subsidy has been tied to the physical unit and recipients benefit only if they consume the services provided by a subsidized producer.

A brief review of federal housing policy follows in order to show the gradual evolution of the delivery system and the basic principles which have shaped its development.

FEDERAL INTERVENTION

The federal government did not become involved in housing until the depressed conditions of the 1930s created the political climate conducive to federal intervention. The major pieces of legislation are described below:[2]

- The initial involvement of the federal government in housing was to create the Federal Home Loan Bank Board and twelve district banks in 1932. The major purpose of the Federal Home Loan Bank System was and still is to provide savings and loan associations with a source of funds to meet seasonal needs as well as to provide liquidity when credit market conditions become tight.
- A second effort was made to ensure a supply of loanable funds for housing in 1933 when Congress established the Federal Deposit Insurance Corporation to insure deposits in commercial banks and the following year the Federal Savings and Loan Insurance Corporation to provide a similar service for savings and loan associations. Federal insurance lowered the risk to potential depositors and thereby encouraged additional deposits which could subsequently be available for loans.
- Mortgage markets were further strengthened by the passage of the National Housing Act of 1934 which created the Federal Housing Administration (FHA) as an independent federal agency for the purpose of insuring residential mortgages. The mortgage insurance program set a ceiling on the rate of interest and provided longer maturities and higher loan to value ratios than were available from private financial institutions. (A similar program, VA, was made available to veterans in 1944 by the Veterans Administration.) The primary benefit of the FHA, later the VA, was that it provided the institutional mechanism to stimulate the private sector to lend more (lower down payment) for a longer period of time. Later in the 1960s FHA insurance was extended to cover the financing of subsidized housing by private financial institutions, and it was only by reducing the risk to the private sector through governmental insurance of the entire loan that private financing could be obtained.

- Prior to 1937, federal legislation was directed toward stabilizing financial markets. The Housing Act of 1937, however, legislated the entry of the federal government directly into the problem of providing housing for the poor. Although changes have occurred since its inception, federal involvement included the following: (1) an annual subsidy to cover debt service on the bonds issued by local housing authorities to finance the construction of the units, (2) special subsidies for the elderly and the handicapped, and (3) additional subsidies for operating expenses. Under the public housing program, the federal government created the financial mechanism for providing housing to the poor but did not involve itself in the construction or management of the project. Local communities through their public housing authorities decided how many units to build and made locational and administrative decisions. Federal responsibility included reviewing the applications and the proposed rent schedules. Under the conventional public housing program, the local housing authority selected the site, designed the project, and then solicited bids for construction. In order to decrease processing time, a "turnkey" method of development was devised in the late 1960s. Under this program, site selection and design is left up to the builder. The housing authority specifies only the critical features of the development, such as number of units and rooms, and then requests bids.
- Mortgage market liquidity was strengthened with the creation of the Federal National Mortgage Association (FNMA) in 1938 to provide a secondary mortgage market for home mortgages. FNMA was originally authorized to buy and sell only FHA-insured and VA-guaranteed mortgages, however, in 1970 its activities gradually were enlarged to include conventional mortgages as well. In 1968 FNMA was converted into a private institution and the Government National Mortgage Association (GNMA) was created at that time to deal exclusively in federally subsidized mortgages. To provide additional mortgage market liquidity, the federal government sponsored the creation of the Federal Home Loan Mortgage Corporation (FHLMC) in 1970 to help establish a secondary market in conventional mortgages.
- The Housing Act of 1949 called for "the realization as soon as possible of the goal of a decent home and a suitable living environment for every American family." Title I of the Act provided for federal aid to local government for urban renewal projects. The Act was significant because it made a commitment to upgrade housing and remove slum conditions on a large scale. However, the congressional support required to finance significantly higher levels of activity was never forthcoming, and fewer than 500,000 units of additional public housing were built prior to the passage of the 1968 Act.

- In 1959 the federal government broadened its housing involvement by creating a program to meet the housing needs of the elderly by making mortgage financing available on preferable terms. The program (Section 202) provided direct loans at a 3% rate of interest payable over 50 years to nonprofit sponsors. In 1961 a subsidized below-market interest rate program was established to provide housing for moderate income families [Section 221(d)(3)]. The term of the mortgage was set at 40 years and the rate of interest was fixed at 3% in 1965. These programs involved direct loans from the federal government and hence had a large, immediate budgetary impact.
- In 1965 a leasing program was added to the public housing program (Section 23). It gave the housing authority more flexibility in its program by permitting existing units to be leased from the private sector. Hence, through this approach public housing tenants could be spread throughout the locality and the stigma of living in a public housing project was minimized.
- In 1965 a rent supplement program was also established. It was designed to provide a mechanism whereby the private sector—through nonprofits, limited dividends (a profit-making corporation or partnership with a limited cash return), or cooperatives—could provide housing for low-income individuals. The income limits coincided with those of the public housing authority. The tenants had to pay the greater of 30% of the market rent or 25% of their income. The balance was subsidized. This program offered a tool to achieve economic integration if it was used in conjunction with some of the moderate-income interest rate subsidy programs described below.
- In 1965 the U.S. Department of Housing and Urban Development (HUD) was created, thereby elevating housing to cabinet-level status. The department was given control over all housing and community development related programming, including FHA.
- The Housing and Urban Development Act of 1968 reaffirmed the goals of the 1949 Act and for the first time specified production targets for both subsidized and nonsubsidized housing for the ensuing decade. The two primary provisions of the 1968 Housing Act included Section 235 (homeownership) and Section 236 (rental). These sections provided greater interest rate subsidies than were available under the previous programs (subsidized down to as low as 1%); utilized the private sector to provide the mortgage money (the federal budgetary impact was just the amount of the annual subsidy); and varied the amount of the subsidy based upon the recipient's income (if the income changed so did the subsidy). Sponsors included nonprofit, limited dividend, and cooperative organizations. Nonprofits or cooperatives received mortgages up to 100% of project costs, whereas limited dividend partnerships

were only eligible to borrow up to 90%. However, limited dividend sponsors could apply certain fees and overhead and profit allowances to their equity requirement. Therefore, the cash payment was reduced substantially below 10%. Income limits under both programs were based upon 135% of the public housing authority's upper-income limits for the area. The programs applied to both new and rehabilitated housing. Significant production occurred under this Act, and through fiscal 1975 in excess of one million units of Section 235 and Section 236 were newly constructed or rehabilitated.

- Tight credit market conditions (high interest rates and limited availability of mortgage money) in 1969 precipitated another form of housing assistance, the "tandem plan." This is an arrangement between GNMA and FNMA which enables GNMA to purchase an FHA-insured or VA-guaranteed mortgage from a lending institution at par or a slight discount. GNMA then sells the mortgage to FNMA at the prevailing market price. The difference between the purchase price and the selling price (the discount) is absorbed by GNMA. Because there are specified ceilings on the maximum interest rates which can be paid on FHA or VA mortgages, discounts must be paid to the lender when market interest rates exceed the specified limits in order for the mortgage to be financed. The "tandem plan" provides a mechanism to reduce the amount of discount the builder or other seller of a home must pay to the lender, it enables mortgage loans to be made at lower rates of interest than would otherwise prevail, and it increases the availability of private mortgage financing. The "tandem plan" originally applied only to federally subsidized mortgages but was extended to most other FHA-insured or VA-guaranteed mortgages in 1971 and to conventional mortgages in 1974.
- In 1969 the Brooke Amendment was passed. This specified that public housing tenants could not pay in excess of 25% of their income for rent. Although operating subsidies in limited amounts had heretofore been available for the elderly and the handicapped, this necessitated operating subsidies on a wholesale basis in order to keep the housing authorities solvent.
- The Housing Act of 1970 authorized an experimental housing allowance program. The experiments started in 1973 and were designed to analyze the response of participants (demand), producers (supply), and alternative administrative mechanisms. Conclusive results will not be available until approximately 1979.
- In January, 1973 a moratorium was declared on all subsidized housing programs.
- In August, 1974 the Housing and Community Development Act was passed which provided for a leased housing program (Section 8). Eligible

sponsors include private developers organized as nonprofits or limited dividends, cooperatives, and public agencies. Financing is provided through private financial institutions, state housing finance agencies, or other public agencies. FHA insurance is not currently available unless a developer qualifies for an existing insurance program. The subsidy is provided by the federal government under a leasing arrangement. The Act also consolidated funding for urban renewal, model cities, water and sewer, and other categorical grant programs and made funds available to state and local government on a formula basis. Local government must submit to HUD a community development and housing assistance plan, but is given wide discretion in the expenditure of funds as long as the activities are designed to meet community development needs and objectives.

- In October, 1975 the suspended Section 235 program was reactivated. The revised program subsidizes the mortgage interest rate down to 5% instead of 1% and is directed at a higher-income household (applicants may have adjusted income up to 80% of the area's median family income instead of 135% of the public housing maximum).

Federal housing policy has stressed private sector involvement. The programs have emphasized strengthening of the primary and secondary mortgage markets, the use of private financial institutions for mortgage money, private contractors and builders for construction and private sponsors (nonprofits, limited dividend corporations, or cooperatives) for locational decisions. Thus many key decisions of program implementation are vested in the private sector.

The federal government has also upheld the autonomy of local government. The federal government has played an advocacy role only to the extent that it has made funds (or matching funds) available for specific programs. The responsibility for taking advantage of the program has been left up to local government, local housing authorities, or the private sector.

Most of the federal programs have been building programs. Use of the existing stock has received lower priority. In addition, most of the federal programs have not included social service components. Individual or family problems related to economic deprivation have not been taken into account in housing legislation even though many of those served bring these problems with them.

The output of these programs in terms of units constructed, leased, or rehabilitated has been small in terms of the need.[3] Approximately 3.2 million units have been assisted since the 1930s (U.S. Department of Housing and Urban Development, 1974b: 575), two-thirds of these occurring since 1968. The primary method of providing housing to the poor has thus been through the filtering process. The major federal housing commitment has been the amount of federal tax revenue foregone as a result of the treatment of mortgage interest and property taxes as expense items in computing federal tax liabilities, an amount that annually has exceeded the total budget of the U.S. Department of Housing and Urban Development.[4]

The programs have served the poor and nonpoor alike. Approximately 40% of the assisted units have served moderate-income households. The "tandem plan" and efforts made to strengthen financial markets have benefited directly only those households that can afford to own a home.

DELIVERY SYSTEM

The delivery system for federally assisted housing, depicted in Figure 1, consists of every entity that has significant control of or influence over the resulting output. Congress passes legislation, HUD administers the program and provides mortgage insurance, local developers (including public housing authorities) make locational decisions, and private financial institutions and state housing finance agencies make construction and mortgage money available. These entities, acting together, make the key decisions concerning the location, the number of units built, and the clientele served.

The quality of the housing services provided is determined not only by the above but also by local municipalities and agencies which supply municipal and housing related services to the units or the neighborhood in which they are located. In addition, the management entity plays an important role in determining overall tenant satisfaction. The output of this delivery system and the evaluation thereof is dependent upon the incentives which influence the decision-making calculus of each of the principal components of the system. A discussion of these follows.

CONGRESS

Special interest groups play an important role in the development of national housing policy.[5] These include mortgage bankers, financial institutions, labor, real estate organizations, builders,

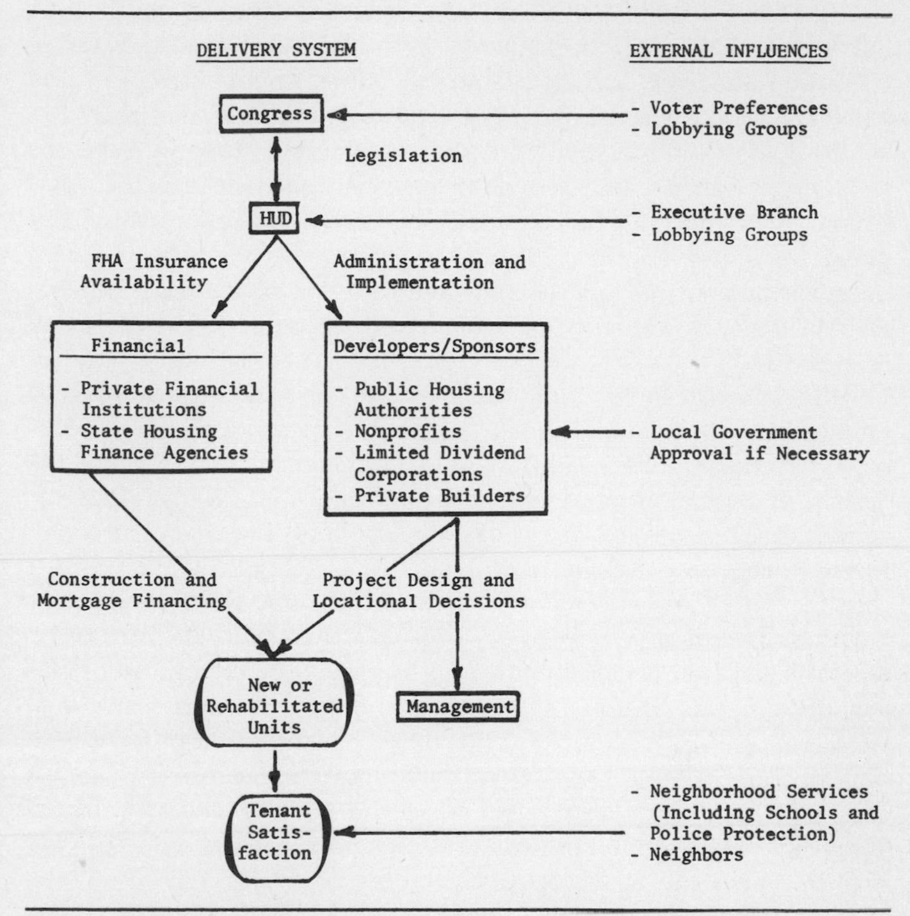

Figure 1: DELIVERY SYSTEM FOR FEDERALLY ASSISTED HOUSING AND EXTERNAL INFLUENCES

building products organizations, and related entities. These groups can expect a direct payoff from legislative action, organization costs are usually not large, and dues pay the ongoing expenses; therefore, the ability of the group to be effective over a long period of time is much greater than that of an organization with a diverse set of interests and no long-term agenda.

Politicians have an incentive to increase their chances of reelection and hence will be motivated to support programs desired by those who can enhance their election opportunities (Downs, 1957: chapters 2, 4). A strong organized lobbying group will thus be able to exert undue influence, not only because it can supply resources for the reelection of the politician but also because it may be the only voice heard on a particular subject (or at least the most vocal one).

If the costs of organizing a group or collecting dues are large, then that particular group may never form. For instance, tenants of subsidized housing are scattered throughout the country, and therefore the costs which would be required to organize them into an effective political force would be substantial. In addition, the resources of such a group would be small, and hence a sustained course of action may not be possible; therefore, the impact of this group on housing policy would be slight. Because the costs of organization are large and the expected gains are small, this group has had difficulty working as an effective lobby agent for a sustained time period.[6]

Politicians thus will have an incentive to listen to the special interest groups and to formulate legislation incorporating the needs of those groups. Federal programs may therefore reflect more the desires of special interest groups, which are already organized for action, than they might the needs of those most intimately affected. Hence, needs are usually defined by the nonpoor and may not coincide with those as perceived by the target population.

Other potential inefficiencies arising from the way in which federal legislation is developed occur as a result of the multiple goals and the myriad different programs. Lobbying groups have had various goals and objectives, and these have been translated into housing legislation which attempts to not only provide standard assisted housing but also to stabilize the housing industry, retard decay of the central city, obtain higher wages for construction workers, promote equal opportunity, and improve environmental quality. Many of these objectives conflict. For instance, urban renewal has resulted in a net decrease in housing for low- and moderate-income groups; also, higher construction wages resulting from the Davis-Bacon Act[7] requirement—that prevailing wages be paid to workers on federally financed or supported projects—and regulations concerning environmental quality have raised the cost of new construction, thereby serving as a deterrent to production.

The multiple objectives of federal housing policy, each having its own set of constituents, mean that changes in the basic nature of federal housing programming may be difficult to achieve. Moving from a direct construction approach to a housing allowance program, for instance, would most likely be opposed by the well-entrenched lobbying groups. Since beneficiaries of such a change, prospective tenants, are not well organized, it would take strong public pressure to accomplish a significant reorientation of federal housing policy.

The congressional component of the delivery system is difficult to mobilize and slow to respond. This must be recognized in formulating alternative solutions to the housing problem or in recommending changes in the delivery system. Although theory may indicate that a change in policy is desirable, unless broad-based support is mobilized, new policies or programs will not be forthcoming. Change, therefore, may best be accomplished incrementally through alterations in the existing set of institutional arrangements in order to make the system perform more efficiently.

Housing expenditures have received low budgetary priority in the past, and there is little reason to expect this to change in the future. Hence, new programs or changes in existing ones will most likely not receive a large increased budgetary allocation. Furthermore, unless the housing allowance experiments produce significantly positive results, there is no reason to believe that past program approaches will be abandoned in the future.

U.S. DEPARTMENT OF HOUSING AND URBAN DEVELOPMENT

The U.S. Department of Housing and Urban Development (HUD) has been delegated the responsibility for program administration and implementation. HUD must approve project applications, and therefore it has the capability to influence the delivery of housing services through its review process and in the allocation of units among various sections of the country. Biases and objectives of HUD officials may play an important role in determining the manner in which programs are administered. For instance, HUD could control project location (white or black neighborhood, inner city or suburbia) or clientele served (elderly or families) by only approving applications meeting certain criteria. HUD does not, however, originate project applications, and hence its control is limited to the types of applications that it receives from the private sector and local public housing authorities.

Theory on bureaucratic behavior suggests that bureaucrats maximize a complex set of goals and objectives.[8] Prestige among peers, professionalism, budget size, number of services offered, and minimizing conflict or problems resulting from deviation from established rules or norms of behavior are a few possible objectives. Cost minimization, innovation, and changes in the traditional way of doing business may not be ascribed to by bureaucratic producers unless these considerations contribute to the bureaucrat's goals.

Likewise, in the absence of strong external pressure, the agency may not be responsive to its clientele if such responsiveness necessitates a change in established behavioral patterns.

Theory suggests the following hypotheses: (1) bureaucratic agencies will resist change, (2) bureaucratic agencies will not respond quickly to changes in the traditional ways of doing business, and (3) bureaucratic agencies will develop rules and regulations and will follow these religiously.

While these hypotheses are rather obvious, they are important to understand in order to formulate policy that will be implemented in a manner consistent with its original intentions. For instance, an agency's resistance to change is another obstacle to overcome in effectuating new approaches to housing policy. A housing allowance approach, for example, may not only be opposed by the previously discussed vested interest groups but it may also encounter the opposition of HUD bureaucrats. This means that proponents of change will have to overcome additional obstructions, which will necessarily slow down the pace of new program implementation.

The difficulties encountered by bureaucracy in responding quickly to change is illustrated by the problems experienced in the implementation of the interest rate subsidy programs legislated in the 1968 Housing Act. The Act significantly increased HUD's mandate for production by establishing a 10-year production goal of 26 million units, 6 million of which were to be subsidized. This altered the pace of federally assisted housing production, which rose from an average of approximately 55,000 units per year during the period 1961-1967 to over 400,000 units per year by 1970.

HUD was unable to administer the rapid growth. Field personnel were too few, short cuts were taken in the review process, controls were lacking, and fraud occurred.[9] HUD's difficulties were compounded by the fact that FHA regulations were changed and new interpretations were not concisely communicated to those responsible for implementation. The underwriting criterion of economic soundness was waived for special-purpose federal mortgage insurance programs in order to encourage the use of these programs in declining inner-city neighborhoods. The vague term of acceptable risk was substituted for economic soundness, and the criterion for calculating the maximum value of the structure was loosened. As underwriting standards were lowered, higher risk loans were insured, and the probability of default increased.

The adverse publicity which occurred as cases of fraud and rising defaults were reported in many inner-city locations could have been predicted. The problem arose from trying to accomplish social change rapidly through an organization that was accustomed to living by a well-defined set of rules. As the rules were bent and the organization was placed under a political mandate to respond, administration could be expected to break down. The result is not unexpected and could have been administratively corrected given time. The unfortunate consequence was that the publicity provided popular support for the decision to curtail the programs because they were not "workable."[10] The lesson to be learned from this experience is that change must be introduced slowly in order to ensure that it is properly administered, otherwise program opponents will have an issue around which to organize.

The final hypothesis concerns the desire of bureaucracy to make its job administratively easier by organizing rules and regulations to guide decision-making activity. This minimizes the need for individuals to expose themselves and risk their jobs by making unpopular decisions. This also has the unfortunate consequence of making programs less efficient from both a demand and supply standpoint.

The HUD review process and the extensive red tape have most likely served as obstacles to participation in the programs. To the extent that this has occurred, HUD has had fewer applications to choose among in implementing its programs. With choice reduced, there is greater likelihood that the resulting assisted housing will offer tenants fewer alternatives than would otherwise be the case.

Choice is also restricted by the rigidity imposed by the guidelines themselves. In order to maximize the ease with which decisions are made, flexibility in the interpretation of rules and regulations is minimized. For example, HUD imposes maximum mortgage ceilings. Although there is some variation in the permissible maximums among different sections of the country, the result has been to discriminate against subsidized housing in higher-cost areas.

Government regulations have contributed to higher costs of supply: FHA minimum property standards are higher in some cases than those prevailing in the private housing market; Davis-Bacon prevailing wage rate requirements raise the wages of some trades above those paid in private construction; and long review times and FHA inspection requirements contribute to higher costs. In addition, there is little incentive for a producer to minimize production costs because applications are not regarded any more favorably if per-unit developmental costs are below the maximum permissible levels.

The analysis of HUD's rule in the delivery system leads to the conclusion that the existing housing programs are inherently inefficient. This was supported by HUD's own analysis which followed the 1973 moratorium (U.S. Department of Housing and Urban Development, 1974a: chapter 4). The in-house evaluation found the programs to be inefficient from both a supply and a demand standpoint. Total development costs were higher than for corresponding units produced in the private, nonsubsidized housing market. From a demand standpoint, tenants of the units valued the subsidy less than the real resource cost to government.

These conclusions do not necessarily obviate the need for the programs. Social benefits are created that may more than offset these inefficiencies. However, given that inefficiencies are prevalent, the question for a policy maker to consider is how to reduce the magnitude of the inefficiencies. This question will be explored later.

At this point in the analysis of the delivery system it is necessary to emphasize that the biases and objectives of the HUD bureaucracy, shaped by over 40 years of program administration, are not apt to change dramatically in the future. This underscores the practicality of working toward change gradually without drastically altering the existing set of institutional arrangements.

PUBLIC HOUSING AUTHORITIES

Public housing authorities are created by state or local units of government to own and manage assisted housing for low-income households. Most of the financing for public housing projects has been provided by the sale of tax-exempt securities secured by an annual contributions contract of the federal government. The authority manages directly or subcontracts out to a private firm the management of its units.

Public housing authorities are constrained by political, financial, and market influences. However, in most instances political constraints are overriding. Members of the board of directors are appointed by local elected officials for a limited period of time, and the executive director is appointed by the board. Decisions of the authority would therefore be expected to be sensitive to political considerations.

The importance of political concerns is evidenced by the decisions of some authorities not to build outside of ghetto neighborhoods and to utilize admission policies that perpetuate established racial patterns.[11] A study of locational decisions of public housing authorities within the Pittsburgh Metropolitan Area provides empirical evidence to show that building occurred primarily in the poorest sections of the housing market (Ahlbrandt, 1974). Thus, racial and economic integration may not be achievable through this developmental mechanism.

Financial constraints on housing authorities are imposed by HUD and Congress. HUD approves the developmental costs as well as the yearly operating budgets, rent levels, and amount of operating subsidy available, if any. Since housing managers are rewarded on the basis of the number of units managed and size of the budget, they may not be motivated to minimize costs as long as rental income is sufficient to cover operating expenses and debt service or as long as the deficit is covered by a governmental subsidy.

Innovation in the delivery of public housing services may also be discouraged. To the extent that innovations require new procedures, there may be internal resistance; likewise, if failure of an experimental program would discredit a manager, there is also a disincentive to innovate.

The authority has an incentive to avoid making the management task difficult. This translates both into an admissions policy that caters to tenants who are most likely to pay rent on time and who are the least likely to have social problems and into a building policy which concentrates upon the elderly. (In the Pittsburgh Metropolitan Area, 82% of all units built during the period 1964-1973 were for the elderly.)

Public housing authorities have a significant amount of discretion in program administration. In maximizing its own objectives, the authority is likely to discriminate against those with the greatest housing needs (lowest-income households and those having significant social problems). The private sector is asked to house these individuals. Therefore, to the extent that the objectives of national housing policy run counter to local interests, there is a high probability that national goals will be subverted to local objectives. There is no incentive for HUD to get involved in this matter as long as its regulations are not being violated in the process.[12]

The unresponsiveness of this developmental vehicle to the needs of its clientele is poignantly illustrated in the extreme by the St. Louis Public Housing Authority's demolition of a large portion of its Pruitt-Igoe public housing complex because tenants refused to live there (Meehan, 1975). In Pruitt-Igoe and other projects throughout the country, errors in the design (units too small, population density too high), construction (poor quality), or location (in the midst of a deteriorating community, removed from job opportunities) and the inability of management to cope with the social problems of the tenants (lack of resources or desire) has helped to create undesirable living environments that are shunned by those with alternative housing choices. The implications of the poor planning decisions are not communicated until rising vacancy rates, increased rent delinquency, and a higher incidence of vandalism makes the project economically a failure. This is not meant to imply that all public housing is undesirable. On the contrary, a large portion is excellent housing. However, the Pruitt-Igoe example does illustrate the potential results of a delivery system which stresses producer biases rather than consumer preferences.

NONPROFIT SPONSORS[13]

Nonprofit sponsors (e.g., neighborhood groups or churches) have been eligible to develop housing under the subsidized programs since 1959. Under the interest rate subsidy programs, political constraints were not as great for a nonprofit as they were for a public housing authority. Unless zoning changes were required or rent supplement units were to be included in the development, nonprofits were not required to obtain the approval of the municipality. (Now under Section 8 all projects must be approved by local government.)

Use of a nonprofit for development under the interest rate subsidy programs had both negative and positive ramifications. On the negative side, it enabled groups with little capital to sponsor housing. Unfortunately, many of these groups lacked management expertise and were unable to manage effectively, and the projects defaulted on their mortgages and went into foreclosure. (HUD could have screened out the applications which did not include an experienced management agent.)

On the positive side, nonprofits had a myriad of different goals and objectives. Many times these were neighborhood or church organizations that wanted to provide decent housing for a specified clientele. These groups identified closely with the target population and attempted to provide the housing desired by that group. Naturally, project location usually corresponded to the community in which the group was active. In these cases the preferences of the clientele were more likely to be considered in the development of the project.

Nonprofits in the Pittsburgh Metropolitan Area have sponsored half of all federally assisted housing, other than public housing. Nonprofits accounted for most of the rehabilitated units, the majority of which are in the poorest sections of the city, areas avoided by profit-making developers (Ahlbrandt, 1974). Nonprofits saw a need which was not being met in these areas, and the rehabilitated housing contributed to an upgrading of the neighborhood.

The nonprofit is a mechanism through which housing needs in older, higher risk neighborhoods can be met. Disadvantages of this vehicle can be overcome through the utilization of experienced management firms.

LIMITED DIVIDEND SPONSORS

Limited dividend sponsors are profit-making entities with an annual cash return limited to 6% of the equity investment. Since the cash return is restricted, the primary attraction of this developmental vehicle to investors is to provide them with a tax shelter (Wallace, 1972). Rents are controlled by HUD, and therefore the investment strategy is to minimize the variation of income and expenses around expected levels. This risk minimization strategy leads the investors to consider relatively safe projects. This then translates into serving a nonproblem population in a low-risk location, e.g., the elderly in less deteriorated inner-city neighborhoods or suburban communities.

There is no assurance that housing will be developed in areas where the need is the greatest. In fact, there is a strong indication that this developmental vehicle will not provide housing in poorer neighborhoods, neighborhoods served by the nonprofits. In the Pittsburgh Metropolitan Area most of the activity of limited dividend sponsors was outside of the city limits in stable neighborhoods: median family income only slightly less than the countywide median, high homeownership, low percentage of minority population (Ahlbrandt, 1974).

This form of a developmental vehicle does not have an incentive to minimize costs. Depreciation, and hence the amount of income sheltered, is directly related to total development costs. In addition, developers operating under either a nonprofit or limited dividend status are not motivated to reduce costs because the builders' profits and fees are based upon a fixed percentage of costs incurred.

The limited dividend model may never be an effective mechanism for providing housing in higher-risk sections of a metropolitan area as long as the return to the developer is the same as for building in less risky areas. Although this mechanism may encourage social and economic integration, it does not contribute to a strengthening of core inner-city neighborhoods.

PRIVATE BUILDERS

Private home builders became a part of the delivery system for federally assisted housing under the Section 235 program in 1968 (the program provided an interest rate subsidy for moderate-income households purchasing new or substantially rehabilitated housing). The objective of builders is to maximize profits, and hence they will build according to their perceptions of current market conditions. A study of building activity in the Pittsburgh Metropolitan Area showed that the single family Section 235 units are widely dispersed and are located in areas with incomes slightly in excess of the county median (Ahlbrandt, 1974). Hence, this mechanism was useful in achieving a dispersal of moderate-income families throughout the area but may not be a good device for upgrading housing in poorer communities.

Of all of the programs analyzed by HUD, Section 235 performed the best in terms of both demand and supply efficiencies. From a supply standpoint the units did not cost more to construct than did unsubsidized units (land costs or amenities may differ, however). On the demand side, the recipients valued the subsidy only slightly less than the real resource costs of the subsidy (U.S. Department of Housing and Urban Development, 1974a, chapter 4).

The Section 235 program utilized market forces to a greater extent than did the others, and this most likely explains why it was more efficient than the others. Housing opportunities for a Section 235 buyer were greater in number than for many renters of subsidized developments, and therefore to be competitive builders had less leeway in their construction costs and had to produce units offering similar services to those in the nonsubsidized market.

FINANCIAL INSTITUTIONS

Most of the recent federally assisted housing programs have utilized private financial institutions to underwrite the construction costs and mortgages. Lending policies of financial institutions are normally conservative. The institution's primary objective is to protect the assets of its depositors. As a result, high-risk lending situations are usually avoided unless appropriate security is pledged as collateral for the loan or insurance is available.

Lending criteria utilized by financial institutions take into account the quality of the neighborhood and the direction of neighborhood change (Ahlbrandt and Brophy, 1975b: 26-32). Neighborhoods having a declining economic base are not regarded as desirable lending situations under normal circumstances. Likewise, financing housing developments with a high percentage of low- or moderate-income tenants is not considered a prudent risk in the absence of project insurance because the assurance of having a stable rental income stream is not as great as in developments catering to higher-income tenants.

In order to encourage financial institutions to participate in the federal assistance programs, FHA insurance has been necessary. Most of the insurance programs, however, have provided coverage for the entire loan, alleviating the financial institution of all risk. This has been successful in obtaining private financing, but it diminished the need for financial institutions to analyze the soundness of the investment. This helps to explain some of the defaults and scandals which occurred in the early 1970s. HUD did not have the machinery and the financial institutions did not have the incentive to perform an effective screening function. A partial loan guarantee may have produced different results. Certainly as the exposure of financial institutions increases, their incentive to review the soundness of the loan becomes greater. At some point, however, funds for projects in higher-risk neighborhoods would no longer be available. The point at which that would occur is worthy of additional research.

MANAGEMENT

As discussed earlier, tenants' satisfaction with housing depends upon many factors in addition to the unit. Obviously, the locational decision is an important determinant of the quality of the neighborhood surrounding the unit and directly influences satisfaction with the services derived from the dwelling. Many recent studies have stressed the importance of management (Ahlbrandt and Brophy, 1975a; Sadacca et al., 1974).

Management may be provided directly by the owner of the development or it may be provided by a private management firm under contract to the owner. The incentive of a private management firm is to maximize gross rental income because the fee is usually a percentage of the rents collected. Unless tenants pressure management to provide certain services or to perform in a stipulated manner, and unless the management agent agrees that it is in its best interest to accommodate the wishes of the tenants, there is no assurance that management will operate in a manner consistent with tenant preferences.

Management style has been shown to be an important ingredient in determining tenant satisfaction, and tenant satisfaction has been shown to correlate positively with the financial performance of the development; therefore, management is able to influence the financial viability of the development through the manner in which it interacts with the tenants. Also, management decisions regarding admissions policy and maintenance priorities influence tenant satisfaction. Hence, the greater the tenant input into the management function, the more probable the outcome will be commensurate with the desires of the tenants.

INSTITUTIONS–SUMMARY

The above discussion highlights the constraints affecting the delivery system. Each institution operates under a set of biases and objectives that influence its degree of participation and the manner in which the objectives of federal housing policy are implemented. The various sponsors discussed utilized different locational criteria and were concerned with housing clientele of different socioeconomic characteristics. The appropriate set of institutional arrangements therefore depends upon the desired outcome.

The delivery system is not designed to be responsive to a nonorganized, nonvoting clientele and, as a result, demand inefficiencies should prevail. All of the subsidized programs studied by HUD were deemed inefficient from the standpoint that the services provided were valued less than their real resource cost. Accordingly, increasing the influence of the consumer in the decision-making process will reduce this form of inefficiency. Likewise, most of the programs were also found to be inefficient from a production standpoint. This could be reduced if emphasis were placed upon cost competition among project applications or if greater use were made of existing stock of housing rather than new or substantially rehabilitated units.

Changes in housing legislation or in the institutional arrangements under which federally assisted housing is provided may not be expected to occur rapidly. If anything, the problems encountered in the great production push of the 1968 Act and the failure of the industrialized housing concept to significantly lower production costs may thwart new innovations for the time being.

The next section examines alternative institutional arrangements which may improve upon some of the inefficiencies previously discussed. The housing allowance approach is discussed in detail because if any major policy change is to be introduced it will most likely incorporate some form of this program. For reasons presented earlier, however, it is unlikely that this will occur, at least until the results of the experiments are thoroughly analyzed. The section then focuses upon more readily acceptable institutional changes which increase the input of the recipients into the delivery system.

ALTERNATIVE INSTITUTIONAL ARRANGEMENTS

HOUSING ALLOWANCES[14]

Most of the federally assisted housing programs attach the subsidy directly to the physical unit.[15] Locational decisions are made by the producer. From the consumer's standpoint there is no assurance that the units provided will be in a location that offers a preferred total package of services. However, as long as the subsidized rent is low enough to compensate for an otherwise inferior living environment, the resulting inefficiencies in terms of reduced tenant satisfaction will not be observed. Thus, providing the subsidy through the producer diminishes the amount of information on consumer preferences that would be available if the consumer could apply the subsidy over a larger set of market choices. There is minimum incentive for a potential recipient to voluntarily convey this information to the producer because the likelihood of that individual changing the production decision is negligible.

Tenants' preferences for changes in their housing environment are conveyed, if at all, after the units have been built. This may take the form of declining demand for housing in that project, hence rising vacancy rates, and/or increased maintenance requirements in response to vandalism. Higher levels of dissatisfaction are communicated directly to the owners through physical conditions in the development, contact with tenants, rising operating costs, and reduced cash flow. Remedial solutions may be slow in developing or may not occur at all if the resources are not available.

The articulation of preferences under this system is imprecise and occurs mainly after the fact. A housing allowance program, on the other hand, greatly facilitates demand articulation. Although there may be many variations of the overall approach, in general recipients of the subsidy are given additional income to supplement their housing expenditures. The neighborhood and the services flowing from the unit itself are selected by the tenant. Expanding housing choices, in theory at least, should increase tenant satisfaction from the expenditure of the subsidy monies.

In addition, efficiencies on the supply side should be improved. Producers, competing for the subsidized clientele, should be motivated to offer the services desired by the consumers at a competitive price. Price competition should be limited only to the extent of the availability of alternative sources of supply. Although in the short run producers may be able to capture some of the subsidy in terms of excessive profits because of the difficulty of expanding supply rapidly in response to an increase in effective demand, this will be mitigated in the long run by new production, rehabilitation, or conversion of existing units.

A housing allowance, by increasing the freedom of choice of the consumers, creates a market-like situation, and the expected results should be commensurate with the operation of a freely functioning housing market. Consumers could be expected to spend up to the point where an extra unit of consumption approaches the value of the resources foregone, and producers should produce up to a like point.

Market imperfections could impair the results. Housing market discrimination or an inadequate flow of information concerning the availability of units could reduce potential tenant satisfaction. Likewise, a tight housing market coupled with little response in terms of new construction could siphon off a large portion of the subsidy to owners with little resulting increase in quality.

Although the experiment in theory seems to offer an improvement in the delivery of housing services, in practice it will only work if controls are established to protect the recipients from exploitation and if the program is funded at an adequate level for the recipients to purchase standard quality housing. The latter may not occur. The federal government, through its welfare programs, is already funding a housing allowance program of sorts. Welfare payments are low, and in many states the recipients are forced to live in substandard housing (U.S. Department of Housing and Urban Development, 1974a: 46-48).

If a housing allowance program is adopted, support for housing assistance among the general electorate may be diminished as special interest groups turn their attention elsewhere, and hence program funding may be inadequate. This is a consideration which must be taken into account when new legislation is being developed. However, the likelihood of a major policy change before the end of this decade is remote. The next section, therefore, focuses upon more immediate remedies to current and potential program inefficiencies.

MANAGEMENT/OWNERSHIP

Changes in the tenant/management relationship may not only increase tenant satisfaction with the services provided but may also enhance the financial viability of the development. New institutional arrangements may be implemented by giving tenants greater control over management decisions or by taking an additional step and transferring ownership to the tenants through a housing cooperative. The theory is that as management comes under the control of the tenants, management will be more responsive to the needs of those residing in the development.

Management influences tenant satisfaction through its tenant selection, rule enforcement, and maintenance policies as well as the manner in which it interrelates with the tenants and responds to their needs and requests. A private management firm has a wide degree of latitude in which to operate and may not necessarily provide the services desired by the tenants. As long as the tenants do not articulate their preferences to the management in a manner that forces the firm to change its policies, tenant satisfaction will be lower than would otherwise be the case.

As tenant control over management decisions increases, there is a greater incentive for residents to become involved in the decision-making process and to articulate their preferences because their input may be observed to make a difference. Likewise, if tenants assume ownership of the development by forming a cooperative with each tenant owning a share, their vested interest in the project's performance increases. There is reason to believe that the financial performance may also improve as tenants become more concerned about excessive expenditures. Peer group pressure may be used to control against vandalism and excessive maintenance requirements as well as to pressure delinquent tenants into paying their rent. Finally, maintenance expenditures may be reduced through the sweat equity provided by the tenants themselves.

Involving tenants in the management of their development is not a new concept. In 1968 OEO funded a tenant management demonstration program in the 38-building (1,216 units) Bromley-Heath project of the Boston Housing Authority. The experiment, originally pilot tested on several buildings, proved to be a success and the Bromley-Heath Tenant Management Corporation was formed as a nonprofit, community development corporation controlled by the residents. The nonprofit entered into an agreement with the housing authority to undertake the management of the entire development. The demonstration has been a success as measured by an improved visual appearance of the development (the tenants helped to remove trash and debris), declining rent delinquency, reduced vacancy rate, and improved tenant satisfaction (Zimmer, 1975: 62-68). The St. Louis Public Housing Authority has also piloted tenant management in four of its properties, and HUD has viewed these as a success and is now going to expand the program to other authorities (Bureau of National Affairs, 1975: 321-322).

Cooperative ownership of a development by tenants may produce similar results. Case study research on the conversion of subsidized rental developments into cooperatives has shown improvements in financial performance and tenant satisfaction. The conversion process, however, needs the following preconditions for success: (1) assistance from experienced professionals, (2) tenant interest and leadership, (3) good management, and (4) adequate cash flow (Zimmer, 1975: 98-102).

Tenant ownership has been shown to make a project stronger financially, but it cannot be expected to work miracles if the development is unable to meet normal operating expenses and debt service given a realistic occupancy rate. Cooperative conversion may be a viable alternative institutional arrangement for assisted projects currently experiencing financial difficulties, but the project must first be put on a sound financial basis prior to conversion. Furthermore, cooperative conversion or tenant control may not produce positive results if the development is located in a neighborhood with an undesirable living environment or if an incompetent management firm is hired.

OPTIONS FOR LOCAL GOVERNMENT

The analysis of the delivery system and the incentives confronting each of the participants provides a starting point to aid in program formulation at the local level. The resources available to local government include its operating, capital improvement, and community development budgets and the utilization of the current federal housing programs.

LOCAL HOUSING PROGRAMS

The nature of housing problems is economic—low income households do not have sufficient income to purchase standard quality housing. Local government, however, is not equipped to deal directly with this cause as it would involve a significant amount of income redistribution, and, for reasons discussed elsewhere (Musgrave, 1959), this is most efficiently performed by the federal government.

Local governments' approach toward housing related programming is more appropriately directed toward stemming neighborhood decline. In the past, the tools available have included urban renewal or a variation thereof, Federally Assisted Code Enforcement, Model Cities, and public housing. Other federally assisted housing programs involving new construction or substantial rehabilitation were under the control of private developers, and locational decisions may not have coincided with the overall neighborhood improvement goals of the local municipality, if in fact these objectives existed. The Housing and Community Development Act of 1974 not only gave local government greater discretion in the expenditures of the block grant funds but also required it to coordinate assisted housing with its overall community development plan. Since local government must approve the locational decisions of private developers, theoretically greater coordination may be achieved.

The housing related programs undertaken by local government will most likely emphasize neighborhood preservation. These programs give elected officials the opportunity to maximize their reelection opportunities by spreading resources around the city immediately rather than undertaking a large, urban-renewal-type project, which concentrates a significant funding effort in a limited area and does not produce a visible return for many years.

Neighborhood preservation programs of local government which are not comprehensive in nature will have little effect. A revolving home repair loan program administered citywide will be too dispersed to produce a visible impact on any one location. An urban homesteading program is capable of impacting only one structure at a time and will have no neighborhood effects unless it is utilized as a part of a much broader strategy aimed at reclaiming a given community. Likewise, a rehabilitation loan program will not be effective unless it is concentrated in a specific area or becomes a part of a larger programming effort.

Local governments' approach toward neighborhood preservation thus needs to be comprehensive. Given that neighborhood residents must want to remain in the neighborhood and that financial institutions must be willing to make loans in the neighborhood for a preservation strategy to be effective, it is important to involve these parties in program development.

Federally assisted housing, both new construction and substantial rehabilitation, must be considered by local government in the formulation of its neighborhood preservation strategy. This is particularly important because of (1) the effect of the neighborhood environment on the quality of the housing services provided, (2) the importance of investment to the quality of the housing stock of a particular community, and (3) the effect of subsidized housing on the investment mentality of those residing in the area (positive and negative implications).

To the extent that it is desirable to encourage the development of federally assisted housing in a neighborhood, it may be necessary for local government to offer additional inducements. Limited dividend sponsors may not have an incentive to develop housing in declining neighborhoods, and unless the risk is reduced or the return increased, development may not occur. In addition, public housing authorities and limited dividend sponsors may not be motivated to build housing for families, preferring to concentrate on the elderly.

AN ALTERNATIVE DELIVERY SYSTEM

Neighborhoods decline because they are no longer able to retain an increasing or at least a stable economic base. For one of many reasons, a neighborhood becomes less competitive within the total housing market and begins to experience a reduction in the effective demand for housing. This results in declining or nonincreasing prices (rents) and lower rates of return on investment.[16]

Maintenance of the housing stock becomes a critical problem.

Financial institutions may no longer be willing to extend credit in the area as their perceptions of risk increase. Even if credit is available, many property owners may not be able to afford it or may be unwilling to upgrade their property because of a negative investment psychology, which is induced by the presence of deteriorating property in the neighborhood, inadequate public services, racial change, undesirable neighbors, and the like. In short, investors may lose confidence in the neighborhood.

Neighborhood preservation programming must take into account the comprehensive nature of the problem. In order to stem decline, the investment psychology of homeowners and investors toward the neighborhood must be altered. This may require different approaches depending upon the specifics of a given neighborhood. However, in general it requires targeting a neighborhood for a home improvement loan program, providing subsidies if necessary, utilizing codes enforcement to ensure that property owners respond, and providing public improvements to show property owners that the city is committed toward preserving the neighborhood. If private financial institutions can be brought into the program, city funds will be leveraged and the prospects for success enhanced.

In the past, federal programs directed toward upgrading inner-city neighborhoods were normally operated by a city agency. Citizen involvement in a decision-making capacity was usually not encouraged. If citizen participation was mandated, the decision-making process was so long that effective input was discouraged. The involvement of the private sector in the programs did not occur because bureaucrats were not accustomed to seeking out the private sector, and extensive governmental red tape posed a further disincentive. In short, the traditional approach toward neighborhood preservation was a one-sided governmental effort.

The lack of success of many of the federal efforts is therefore not surprising. Inadequate incentives were built into the system to encourage the development of comprehensive neighborhood preservation programming, and the needs of those in the neighborhood were often not taken into account by the bureaucrats who formulated the programming. In addition, the orientation of city agencies was more toward direct measures of program performance (e.g., number of code inspections made, amount of money expended) rather than toward the less quantifiable and more difficult to accomplish objective of altering the long-term investment psychology in the neighborhood.

A new institutional arrangement for delivering neighborhood preservation programming was pioneered in Pittsburgh in 1968 (Ahlbrandt and Brophy, 1975b: chapter 9) and is now being implemented nationally by the Urban Reinvestment Task Force, a joint partnership of the Federal Home Loan Bank Board and HUD. The model, called Neighborhood Housing Services (NHS), utilizes a privately funded high-risk revolving loan fund to make home improvement loans to nonbankable homeowners living in a target neighborhood, private financial institutions to make bankable loans in the area and to fund the operating budget, and city government to provide codes enforcement and to upgrade city services in the neighborhood. The organization is a nonprofit corporation, usually controlled by neighborhood residents with a strong input from financial institutions.

The NHS brings together the three principal parties controlling the direction of neighborhood change and involves them in a joint effort to upgrade the neighborhood. The process of involving the key decision makers in each sector is given attention equal to that of ensuring that the necessary commitments are made to enhance success.

A workshop process is used to develop a dialogue between the citizens, financial institution representatives, and local governmental officials. During this time commitments are usually forthcoming. The involvement of each of the parties is predicated upon the participation of the others, and therefore joint agreement and joint definition of neighborhood needs and program alternatives are essential ingredients in obtaining the commitments.

Citizens have an incentive to participate because it is their neighborhood and their input helps to determine the nature of the program and the specifics of city program commitments. The city has an incentive to become involved because of citizen pressure and because financial institutions are willing to support the program, which thereby leverages the impact of city investment in the neighborhood. Financial institutions have an incentive to become involved to protect their existing investments in the neighborhood, to exhibit a social consciousness, to increase deposits, and/or to lessen redlining allegations. In addition, perceived risks to the institution from lending in the neighborhood are lessened as citizen and city commitments toward the neighborhood are demonstrated.

The program is a partnership effort of the three groups, with the city taking a secondary role. The NHS is privately run and privately funded although a portion of the high-risk loan funds may be provided through grants from public agencies or local government. The primary objective of the NHS is to stem decline and to serve the neighborhood. Program statistics are secondary. Although the program is relatively new, results are encouraging (Ahlbrandt and Brophy, 1975b: chapters 6-8; Ahlbrandt et al., 1975).

The NHS program is a new mechanism through which to implement national housing policy. The Urban Reinvestment Task Force's role is to act as a facilitator, helping to bring the program together in interested municipalities. Other than acting as a catalyst and providing technical assistance and perhaps a grant to underwrite a portion of the high risk loan fund, the Task Force's role and hence that of the federal government is minimal.

The program is run locally and program priorities, needs and administrative decisions are established by those involved at the neighborhood level. Red tape is lacking, and, since local participants can become an effective part of the decision-making apparatus, there is an incentive for involvement. Decisions are thus made by those most knowledgeable to do so. Satisfaction of program recipients should be high. In addition, since those responsible for raising and/or providing the funds are also overseeing the expenditures, there is an incentive to control costs. The program therefore has the necessary incentives for its efficient implementation.

The Housing and Community Development Act of 1974 provides local government with significant discretion in the expenditure of the funds. The federally controlled categorical grant programs and the concomitant unresponsive federal administrative machinery may now be replaced by new sets of institutional arrangements. Although the appropriate delivery system may vary depending upon conditions prevailing at the local level and upon the biases of local government, if performance is to be improved the critical consideration becomes one of creating a structure which provides motivation to change established patterns of behavior.

CONCLUSION

This chapter concentrated upon the delivery system for federally assisted housing services. The discussion focused upon federal programming and the institutional constraints and the incentive structure. The delivery system was shown to be unresponsive to the clientele served (resulting in potential demand inefficiencies) and to lack incentives for cost minimization (resulting in higher production costs than for comparable privately produced unsubsidized units).

An alternative solution, which would improve efficiency from both a demand and supply standpoint, is a housing allowance. This form of subsidy replaces a quasi-nonmarket decision-making structure with a market oriented delivery system. The probability of this program being legislated in the near future is remote, and therefore if program efficiency is to be increased, attention must be directed toward making institutional changes within the existing delivery system. Focusing upon the tenant/management relationship and upon the ownership vehicle offers the greatest opportunity for success.

The Housing and Community Development Act of 1974 provides local government with additional discretion in the formulation of housing and community development programming, and, as a result, local governments' role in the housing area has assumed increased importance. Given that local government will most likely concentrate upon neighborhood preservation strategies, it is important that the programming be comprehensive. For maximum effectiveness neighborhoods need to be targeted, home improvement loan funds made available, and public improvements and codes enforcement undertaken.

Incentives confronting financial institutions, citizens, developers, and local government are important determinants of the resulting output. Attention, therefore, needs to be directed toward an understanding of these forces in order to develop strategies incorporating institutional arrangements which maximize the probability of policy objectives being fulfilled. This requires a rethinking of the traditional governmental approach toward housing at all levels of government.

NOTES

1. For a selection of readings on specific dimensions of the housing environment see Ahlbrandt and Brophy (1975a). In addition to those references see Angrist (1974), Onibokun (1974), Sadacca et al. (1974), and Western et al. (1974).

2. For a good overview of federal housing policy see Keith (1973), Schechter (1972) or U.S. Department of Housing and Urban Development (1974a: chapter 1).

3. Housing deprivation in the United States in 1970 was estimated to affect 20% of the nation or approximately 13 million households (Birch et al., 1973: 4-7).

4. Approximately $6.2 billion of federal tax revenue was foregone (U.S. Department of Housing and Urban Development, 1974a: 36), compared to HUD's housing assistance payments of approximately $2.4 billion and HUD's total budget appropriation of $5.3 billion (U.S. Department of Housing and Urban Development, 1974b: 6).

5. For a good discussion of this point see Keith (1973) and Lilley (1971).

6. For a discussion of tenant organizations see Marcuse (1971).

7. The Davis-Bacon Act, passed in 1931, requires the payment of prevailing wages to all construction workers on federally financed or supported projects. The wage rates are determined by the Secretary of Labor and often result in rates considerably above the average for the area.

8. Theory on bureaucratic behavior is presented by Downs (1967), Niskanen (1971) and Tullock (1965).

9. For a discussion of these points see Downs (1973 and 1974) and McFarland (1972).

10. A public opinion survey taken in July, 1973 (U.S. Department of Housing and Urban Development, 1974a: 88) showed strong public support for ending the interest rate subsidy programs. It may be argued that the Nixon Administration chose to curtail the programs not for reasons inherent in the programs themselves but because they were deemed too expensive. President Nixon was concerned with fighting inflation and needed to reduce the size of the federal budget deficit. Expenditures on federally assisted housing doubled during the period 1968 to 1973 and could have been expected to at least double again if the production rates of the early 1970s were continued through 1978.

11. The Chicago Housing Authority is an example of one which concentrated its activity in ghetto areas for political reasons (Yale Law Journal, 1970).

12. An example of HUD's desire to minimize its role in issues not pertaining directly to its regulations is its lack of enforcement of the 1964 Civil Rights Act prohibiting discriminatory action by the federal government and of the 1968 Civil Rights Act requiring the affirmative administration of that provision as it pertained to programs of federal agencies. It was not until HUD was ordered by the court to enforce these provisions (Shannon vs. HUD decision in 1970) that HUD established criteria to implement the intent of the Civil Rights Acts.

13. For a discussion of nonprofits see Keyes (1971).

14. Housing allowances are discussed in Weaver (1975) and U.S. Department of Housing and Urban Development, Office of Policy Development and Research (1975).

15. A good discussion of the inefficiencies resulting from this method of providing a subsidy is contained in Aaron and VonFurstenberg (1971).

16. A detailed discussion of neighborhood decline and revitalization as well as an extensive bibliography on the subject is contained in Ahlbrandt and Brophy (1975b).

CASE

Shannon et al. v. United States Department of Housing and Urban Development (1970) 436 Fed. 2d 809.

REFERENCES

AARON, H. J. and G. M. VonFURSTENBERG (1971) "The inefficiencies of transfers in kind: the case of housing assistance." Western Economic Journal 9 (June): 184-191.

AHLBRANDT, R. S., Jr. (1974) "The delivery system for federally assisted housing services: constraints, locational decisions and policy implications." Land Economics 50 (August): 242-250.

——— and P. C. BROPHY (1975a) "Increasing efficiency in the delivery of federally assisted housing services through an understanding of tenant preferences." Frontiers of Economics 1 (Summer).

——— (1975b) Neighborhood Revitalization: Theory and Practice. Lexington, Mass.: Lexington Books.

——— J. ZIMMER, K. BURMAN, and P. RICHARDSON (1975) The Neighborhood Housing Services Model: A Progress Assessment of the Related Activities of the Urban Reinvestment Task Force. Washington, D.C.: U.S. Department of Housing and Urban Development, Office of Policy Development and Research.

ANGRIST, S. S. (1974) "Dimensions of well-being in public housing families." Environment and Behavior 6 (December): 495-515.

BIRCH, D., R. ATKINSON III, P. L. CLAY, R. P. COLEMAN, B. J. FRIEDEN, A. F. FRIEDLANDER, W. L. PARSONS, L. RAINWATER, and P. V. TEPLITZ (1973) America's Housing Needs: 1970 to 1980. Cambridge, Mass.: M.I.T.-Harvard Joint Center for Urban Studies.

BISH, R. and V. OSTROM (1973) Understanding Urban Government. Washington, D.C.: American Enterprise Institute for Public Policy Research.

BISH, R. and R. WARREN (1972) "Scale and monopoly problems in urban government service." Urban Affairs Quarterly 8 (September): 97-122.

The Bureau of National Affairs, Inc. (1975) Housing and Development Reporter 3 (August 25): 321-322.

DOWNS, A. (1974) "The successes and failures of federal housing policy." Public Interest 34 (Winter): 124-145.

——— (1973) Federal Housing Subsidies: How Are They Working? Lexington, Mass.: Lexington Books.

——— (1967) Inside Bureaucracy. Boston, Mass.: Little, Brown.

——— (1957) An Economic Theory of Democracy. New York: Harper & Row.

KEITH, N. S. (1973) Politics and the Housing Crisis Since 1930. New York: Universe.

KEYES, L. C. (1971) "The role of nonprofit sponsors in the production of housing," pp. 159-183 in papers submitted to Subcommittee on Housing Panels, Part I, Committee on Banking and Currency, House of Representatives, Ninety-Second Congress. Washington, D.C.: U.S. Government Printing Office.

LILLEY, W. III (1971) "Washington pressures/home builders' lobbying skills result in success, 'good-guy' image." National Journal (February 27): 431-445.

MARCUSE, P. (1971) "Goals and limitations: the rise of tenant organizations." Nation (July 19): 50-53.

McFARLAND, C. M. (1972) "Unlearned lessons in the history of federal housing aid." City (Winter): 30-34.

MEEHAN, E. J. (1975) "Looking the gift horse in the mouth: the conventional public housing program in St. Louis." Urban Affairs Quarterly 10 (June): 423-463.

MUSGRAVE, R. A. (1959) The Theory of Public Finance. New York: McGraw-Hill.

NISKANEN, W. A. (1971) Bureaucracy and Representative Government. Chicago: Aldine.

ONIBOKUN, A. G. (1974) "Evaluating consumers' satisfaction with housing: an application of a systems approach." Journal of the American Institute of Planners 40 (May): 189-200.

OSTROM, E. (1972) "Metropolitan reform: propositions derived from two traditions." Social Science Quarterly 53 (December): 474-493.

SADACCA, R., S. B. LOUX, M. L. ISLER, and M. J. DRURY (1974) Management Performance in Public Housing. Washington, D.C.: Urban Institute.

SCHECHTER, H. B. (1972) "Federal housing subsidy programs," pp. 597-630 in The Economics of Federal Subsidy Programs. Joint Economics Committee, Ninety-Second Congress, Second Session. Washington, D.C.: U.S. Government Printing Office.

TULLOCK, G. (1965) The Politics of Bureaucracy: Washington, D.C.: Public Affairs Press.

U.S. Department of Housing and Urban Development (1974a) Housing in the Seventies. Washington, D.C.: U.S. Government Printing Office.

--- (1974b) Appropriations for 1975, Hearings Before a Subcommittee of the Committee on Appropriations, House of Representatives, Ninety-Third Congress, Second Session. Washington, D.C.: U.S. Government Printing Office.

--- , Office of Policy Development and Research (1975) Experimental Housing Allowance Program: Initial Impressions and Findings. Washington, D.C.: U.S. Department of Housing and Urban Development, Office of Policy Development and Research.

WALLACE, J. (1972) "Federal income tax incentives in low and moderate income rental housing," pp. 676-705 in The Economics of Federal Subsidy Programs. Washington, D.C.: U.S. Government Printing Office.

WEAVER, R. C. (1975) "Housing allowances." Land Economics 51 (August): 247-257.

WESTERN, J. S., P. D. WELDON, and T. T. HAUNG (1974) "Housing and satisfaction with environment in Singapore." Journal of the American Institute of Planners 40 (May): 201-208.

Yale Law Journal (1970) "Public housing and urban policy: Gautreaux vs. Chicago housing authority." Yale Law Journal 79: 712-723.

ZIMMER, J. E. (1975) From Rental to Ownership: A Strategy for Improving Low and Moderate Income Housing. Pittsburgh: ACTION-Housing, Inc. (forthcoming [1977] as Sage Professional Paper in Administrative and Policy Studies No. 03-038. Beverly Hills, Calif.: Sage Publications).

2

Juvenile Corrections: A Reform Proposal

DENNIS R. YOUNG
JANET FINK

I. INTRODUCTION

☐ THE LAST DECADE has witnessed major reforms aimed at protecting the due process rights of juvenile offenders in the courtroom.[1] These developments have removed much of the informality of Family Court proceedings. Such informality was originally intended, by the architects of juvenile court, to assist in the rehabilitative process by allowing the judge to function as a substitute parent in correcting wayward youth.[2] The reforms of the last decade redressed some of the legal abuses that emanated from the informality of juvenile court proceedings, but they did not enhance the capability of the juvenile justice system to provide the correctional services needed for juvenile offenders.

Perhaps this was inevitable, for the issue of legal protections is one that is more easily addressed by court decisions and legal mandates. The issue of adequate service provision, however, is one that requires structural organizational solutions. Some states have begun to act along these lines to improve the quality of services,[3] but fundamental solutions are not imminent.

AUTHORS' NOTE: *We would like to thank Judge Arthur Abrams of the Family Court of Suffolk County, New York for his help and cooperation in obtaining information for this paper. The Urban Institute is acknowledged for support in connection with the formulation and data collection phases of the study.*

Structural solutions need to be based on a sophisticated understanding of economic and organizational behavior. In this chapter, we will attempt to develop such an understanding in the juvenile corrections area using a classic paradigm of economics.

In particular, service decisions will be classified into two groups: demand decisions and supply decisions. This may seem to be a rather uncomfortable terminology, at first, for so "un-market-like" fields as justice and corrections. However, the paradigm proves quite useful for scrutinizing the forces influencing correctional decisions and their eventual performance implications.

In the private marketplace, demand decisions represent the consumption choices that individual consumers make among alternative products available to satisfy their needs or preferences. Supply decisions, on the other hand, are the choices made by providers of goods and services (firms) with respect to what shall be produced (type, quality, and quantity of product) and offered to the consuming public. The market system is organized by vesting consumption decisions in sovereign consumers and supply decisions in (multiple, competing) firms. Bargaining in the marketplace is the mechanism that resolves the differences between consumer wants and supplier offerings.

In a corrections system, as with other public services, there is no marketplace, and demand and supply decisions are organized in a different way, but this does not render the terminology any less relevant. Choices are made as to what correctional services offenders will "consume"—e.g., what institutions will house them, what education and training programs they will attend, how they will be cared for and counseled, and when they will be transferred or released. These "demand" choices are vested in public servants and made for the offender-consumer generally without his voluntary consent, but they are made nonetheless within the context of alternatives made available by the system of supply. The demand decisions consist of a series of choices that determine the "trajectories" of individual offenders—from first contact with law enforcement authorities through sentencing, initial disposition, rehabilitation programs and institutions, parole, and ultimate release.

The concept of demand decisions in corrections hinges on the notion of selection among "available alternatives." By implication, therefore, the system of decision-making that leads to supply "offerings" of correctional services is equally important to the determination of the nature of services ultimately rendered. Needless to say, correctional supply decisions are also organized differently from supply decisions in the private marketplace. The determination of how much of what services to offer is made by government bureaus in the context of general government machinery for resource allocation. The actual production and operation of services and facilities may be carried out by the bureaus themselves, farmed out to contractors, or entrusted to private groups. As on the demand side, the decisions involved in formulating correctional supply follow a sequence, and take place at a number of different points in place and time, beginning with budgetary allocations at the state level through program decisions within departmental hierarchies, to service offerings by public institutions, contractors, and other suppliers.

Given the conception of corrections as a system of demand and supply, it is interesting to contrast the structure of the juvenile system with that of the adult correctional system in the United States.

The adult correctional system appears crippled by two salient features: the system is both *monopolistic* and *repressive*. In this system, the offender is turned over to the jurisdiction of a department of corrections following initial sentencing and is essentially removed from the purview of the court thereafter, unless he commits another crime. The department of corrections makes all subsequent decisions regarding offender services. In essence, the supply system (department of corrections) makes its own demand decisions. Thus, consumers of the services, or effective consumer representatives, can neither choose alternative suppliers, because all correctional services are provided by the same governmental agency (department of corrections), nor effectively articulate their needs from within, because they have no economic, administrative, or political leverage to exert. In the terminology developed by Albert O. Hirschman, both the "exit" and "voice" options are closed,[4] hence the adult correctional system is without the necessary "feedback" information and external pressure or incentive to assure effective performance. Adult corrections is a classic case of a tight monopoly that can afford to be impervious to the needs of its clients, except where crises (e.g., riots) bring the system into full public view.[5]

In principle at least, the juvenile correctional system is structured differently; it is not obviously monopolistic or repressive. State training schools and other services provided directly by a state juvenile correctional agency constitute only a part of an array of custodial and therapeutic services available for prescription to the juvenile offender. Furthermore, the interests of the juvenile, or some combination of his interests and those of society at large are, in principle, effectively represented and articulated through the unique role of the juvenile court. In particular, the juvenile court, unlike the adult criminal court, plays an extensive role in selecting (public or private) treatment programs and residential settings, in following up these choices, and in maintaining general responsibility for the welfare of the juvenile throughout his youthful involvement with the criminal justice system. In this sense, the juvenile court approximates the role of a "proxy-consumer" of correctional services for juveniles by making demand decisions independently of the system of supply. By virtue of the economic, administrative, and political resources at its disposal, the court may attempt to exert leverage over that system of supply to induce it to respond to the requirements of its clients as those requirements are interpreted by the court.

On paper, therefore, as viewed via the economic paradigm, the juvenile system would seem to have advantages over the adult system—specifically, pluralistic sources of supply and a strong and explicit mechanism for articulating the needs of its clients. In practice, of course, the juvenile system has not performed ideally as a delivery mechanism for correctional services to juveniles. Failures and abuses abound as they do in the adult system.[6] Nevertheless, a fertile direction for reform may be to effect changes that would bring the juvenile system into closer correspondence to the market-like conception of a pluralistic, possibly competitive, supply system with court-articulated demand, rather than to abandon this idea in favor of some other organizational framework.

It will be the purpose of this chapter to explore the imperfections of the juvenile corrections system from the foregoing "economic organization" viewpoint, with the ultimate purpose of proposing possible avenues of improvement. To do this we shall first consider what prerequisites must be satisfied in order for the juvenile system model to function properly. Then we will look more closely at one component of the system—the juvenile court—to consider whether certain of these prerequisites are now satisfied, or how they could be satisfied by making certain alterations in present arrangements.

II. PREREQUISITES FOR PERFORMANCE OF THE JUVENILE CORRECTIONS SYSTEM

In broad outline, the juvenile corrections system is composed of three important parts: the juvenile court as an institution expressing the demand for juvenile services; a series of public and private institutions and agencies, which together supply the services in response to the foregoing demand; and probation, which serves a multiplicity of functions including information provision and as liaison between the court and the agencies of supply.

The supply agencies are quite varied in type, falling roughly into three categories: (a) programs and institutions provided by state or local government under a department of juvenile, correctional, or social services; these may consist of training schools, reformatories, and certain smaller scale full- or part-time programs within the community and in outlying areas; (b) privately owned and managed agencies that provide residential care or day treatment services and that accept referrals of the juvenile court (for fees paid by the state); and (c) supervision services provided by departments of probation; probation may be an independent (local) government agency or part of the court or government department of juvenile or correctional services.

In order for the juvenile corrections system to function effectively, these various types of suppliers must be sensitive and responsive to the court's demand for services. This requires that supply agencies be motivated, by virtue of their professional, charitable, or financial objectives, to serve the various kinds of juveniles that the court seeks to place. And these nominal objectives must be reinforced by economic, political, or bureaucratic pressures or incentives to encourage the suppliers to respond effectively to court demand. Such incentives may take various forms. For example, supply agency budgets can be tied to the number of juveniles served and the nature of services provided. This is the case, for example, for private schools and institutions that accept juveniles referred by the court and that are compensated on a per diem basis for their services. In order for this incentive system to be effective, supply agencies must be sensitive to financial losses (e.g., depend on current revenues rather than grants or endowments), per diem rates must be competitive with alternative (private) sources of income (patronage) for these agencies, and there must be a sufficient reservoir of supply sources to provide the court with a meaningful choice of alternative suppliers; these prerequisites would prevent supply agencies from charging very high rates, being extremely selective in accepting referrals, or being insensitive to demands for quality improvement or other alterations in service.

Alternative to economic incentives are bureaucratic directives. These result where the court issues direct administrative orders to supply agencies under its control. This would be the case, for example, for probation services reporting directly to judges of the court. This arrangement puts the court directly into the business of administering treatment services and essentially short-circuits the special advantages of the juvenile corrections model of organization, in which demand is organized independently and at loggerheads with supply. Still, the option for the court to provide *part* of the array of treatment services "in-house" may give the court some extra leverage in dealing with external suppliers.

Sources of pressure similar in effect to bureaucratic directives are legal orders wherein the court, by virtue of its legal authority, requires changes in the policies of treatment institutions which it does not otherwise control. Use of the latter mechanisms avoids the "in-house" effects of bureaucratic control, but, because of its extreme nature, cannot be used regularly.[7]

Finally, the political system may be viewed as a source of accountability for treatment suppliers, insofar as court judges wield political influence. This source may be particularly relevant to services provided by governmental departments of juvenile or social services, whose budgets are not explicitly tied to the volume of referrals accepted from the court and which are not directly responsible to the court through formal administrative channels. Hence the basic source of leverage the court may exert over such agencies is to deal directly with legislators or other appropriate public officials. For example, judges testify and lobby with legislatures and bargain with local (county) executives to influence the direction of executive agencies or officials.

In summary, it seems essential to investigate the motivation and reward structure associated with the array of service suppliers in the juvenile correctional system and also to determine how sensitive these agencies are to the demands of the court and what changes might be engineered to enhance their responsiveness. For example, it might simply be that not enough money is available to adequately finance services required by juveniles referred by the court, hence the necessary volume or quality of private supply is not forthcoming. Or it may be that various types of suppliers are financially, politically, and administratively invulnerable to court demands, in which case certain structural changes in the "rules of the game"—such as the policies for determining budgets of public agencies, selecting and advancing certain officials, determining eligibility of private institutions for receiving state payments, and so on—might be considered. Various studies have already made reference to the selectivity of private schools and institutions and the insensitivity of public institutions in serving juveniles referred by the court,[8] but more incisive analysis is required to devise specific solutions which will solve these difficulties.

The present chapter will not delve further into the structure and behavior of the supply institutions, although this seems ultimately necessary. Rather we will present an analysis of the juvenile court to see if it fulfills certain requirements to make it an effective "proxy-consumer" or "demander" of juvenile correctional services. Just as ordinary consumers of private goods and services must have certain motivations, abilities, and resources to make them effective shoppers, juvenile judges must fulfill parallel requirements. However, requirements for judges will be more stringent because they are making consumption decisions for people other than themselves.

In the first place, juvenile court judges must have access to *information* about the services they are "purchasing" for their juvenile charges. Furthermore, they must have adequate knowledge of the juveniles themselves, in order to make customized prescriptions of services.

Second, judges must have the *expertise* to evaluate the information they receive. This is a fairly demanding requirement in view of the technical (behavioral science) nature of the services in question.

Third, judges must have the *resources* at their disposal to take advantage of alternative services available, i.e., they need direct or indirect means of influencing the flow of funds to suppliers that provide services requested by the court. In addition, the court must have sufficient resources at its disposal to allow judges to acquire the knowledge they need about each juvenile and about alternative sources of service supply.

Fourth, the court must have sufficient *alternatives* in terms of the types of services available in order to make meaningful selections for individual juveniles.

Fifth, the court must have sufficient *leverage* over supply agencies in terms of the economic, political, or bureaucratic pressures discussed earlier. If leverage is primarily through financial channels, then the court must have a choice not only in terms of the types of services offered but also in terms of alternative supply agencies to provide these services. Otherwise, the court will be confined to a few supply channels, and the concept of financial leverage via the threat of "taking business elsewhere" will be moot.

A sixth consideration is the question of whether judges face significant *constraints* in exercising their options of selecting and monitoring services for juveniles. For example, in some cases the judges may feel that they do not have full authority to initiate or review action on a juvenile case or, in some instances, the decisions of the judges may be reversed by some other authority. For example, in some states, judges can commit juveniles to a state youth authority for the purpose of training school placement. That authority may decide that a less restraining option is more appropriate.

Seventh, judges must view their *own objectives* in correctional as well as judicial terms. If judges have a restricted view of their own roles as adjudicators of the law, rather than an expanded view as active correctional decision makers, then they will tend to rely on the supply institutions for making correctional decisions and will not exert the necessary selectivity, surveillance, and control over these agencies.

Finally, even if judges do set their own objectives in correctional terms, there must be some assurance that they are conscientious in pursuing these objectives. Hence, it is necessary that judges be *accountable* for their behavior and performance to some source of public and/or professional authority.

In summary, there are a series of requirements centering on information, expertise, resources, alternatives, leverage, constraints, objectives, and accountability that bear on the ability of the juvenile court to serve the functions of demand articulation in the juvenile correctional system. As we will discuss below, the court is assisted in several of these areas by departments of probation. Nevertheless, many of these requirements remain inadequately addressed in present practice.

III. RESEARCH STRATEGY

One can approach the task of gathering information and analyzing the (correctional) decision-making capability of the juvenile court in various ways. Two alternative approaches to such research are (a) to intensively study the operations of the court in a particular locale, through open-ended interviews with judges and detailed observations of juvenile court proceedings or (b) to solicit a wide array of responses from judges in juvenile courts located in different cities and states, through formal written questionnaires. There are, of course, classic examples of successful social science research in both the former (intensive)[9] and latter (extensive)[10] modes. Here we report on a very modest research effort which utilizes elements of both approaches–(limited) direct field observation and personal interviews, as well as a small-scale questionnaire mailed to juvenile court judges. (Within the very limited resources available for our study, it was felt that this dual approach would provide insight into both the variety of practices and the nuances of operation in a given locale.)

Our field observations were made in the Family Court of Suffolk County, New York State, and consisted of interviews with judges and probation officials and courtroom observation on several occasions. The written questionnaire was mailed to 115 judges in five states: California, Illinois, Massachusetts, New York, and Pennsylvania. These states were selected primarily for their similarity in terms of urban/rural composition and size. Respondents were selected from the 1972-1973 Juvenile Court Judges Directory compiled by the National Council of Juvenile Court Judges (1972-1973). The strategy was to obtain an equivalent sample for each state, which provided variety in terms of the sizes of the cities where courts were located. Thus, within each city size category, the same number of solicitations were sent to respondents within each state (see Table 1). Within a city-size/state category respondents were randomly chosen if more than the required number were listed in the directory. The number of solicitations in each category was determined by the number of names in the directory for each category and by the cost of the overall mailing.

Responses were received in June, 1973. The response rate to the mail survey was 20% (23 responses received); higher rates of response emanated from New York and California and from smaller cities, as illustrated in Table 1. Some of the questions asked overlapped those

TABLE 1
MAILED SURVEY SOLICITATIONS AND RESPONSES[a]

	City Size[b]					
State	250,000+	100,000-250,000	50,000-100,000	10,000-50,000	Unknown[c]	Total
California	0/5	0/2	1/5	4/11	1/0	6/23
Illinois	0/5	0/2	1/5	2/11	0/0	3/23
Massachusetts	0/5	0/2	0/5	2/11	0/0	2/23
New York	1/5	0/2	2/5	5/11	0/0	8/23
Pennsylvania	1/5	0/2	1/5	0/11	0/0	2/23
Unknown[c]	0/0	0/0	0/0	0/0	2/0	2/0
Total	2/25	0/10	5/25	13/55	3/0	23/115

a. Entries below the diagonal (/) indicate the number of responses solicited; those above the diagonal indicate responses received.
b. City size where court is located, according to 1970 Census of Population.
c. Respondent failed to indicate location of court.

of an earlier, much more complete and comprehensive survey on the backgrounds of juvenile court judges done by McCune and Skoler (1965) before the great wave of due process decisions for juvenile courts. (We will henceforth refer to that study as the "1965 National Survey.")

IV. RESULTS

By way of caution to the reader, the questionnaire results presented here are not necessarily representative or conclusive because of the small sample size and nonuniform survey response. So, too, our field observations were highly restricted in scope. These limitations notwithstanding, the findings do provide some insights into the efficiency of the juvenile court in fulfilling its demand articulation role in the delivery of correctional services. In the following discussion, findings refer to responses to our mailed questionnaire, unless specifically noted otherwise.

INFORMATION

Judges were asked what kind of information was routinely available in making a disposition decision for a given juvenile offender. All of the 23 judges responding to the mail survey indicated they routinely had information on family background, school record, and criminal record. In addition, all but one judge indicated that some kind of sociopsychological or psychiatric evaluation was available, either routinely or upon request.[11]

Judges were asked to rank their sources of routine information including reports from court intake, probation, schools, police, diagnostic services of the court, juvenile services departments, or independent facilities. Thirteen of the judges ranked probation as first, two judges ranked probation second, and six others checked probation reports as being important but provided no rankings. Every judge surveyed indicated that probation was a source of information at disposition. No other source was as widely used or as highly ranked as probation for dispositional information on the juvenile.

In Suffolk County the probation report constitutes a very important source of information for judges at disposition. This report takes the form of a social history of the juvenile and relies on police, school, and other records, neighborhood and family interviews, and so on, to piece together an understanding of the given child's behavior and motivation for the particular act he is charged with. The probation report goes so far as to recommend a particular disposition alternative to the judge, and, if placement in a school or institution is recommended, the probation department also makes preliminary inquiries to see if openings are available.

Judges in Suffolk indicated that they usually, but not always, follow the recommendations of probation. In addition to probation reports and supplementary mental health evaluations, judges in Suffolk seemed to rely heavily on their own observations of the juvenile in the courtroom to provide them with insight for selecting an appropriate disposition.

When asked what supplementary information they could obtain for disposition if they wanted it, 18 of the judges in the mail survey specifically mentioned psychiatric/psychological reports of some kind. These judges indicated that they requested such reports in a range of circumstances covering violent, emotionally disturbed, and disruptive behavior, and for offenses involving arson, drugs, sex, assaults, runaways, and incorrigibility. The percentages of juvenile cases where these judges requested supplementary reports varied from 1% to 35%. Judges in New York indicated a generally higher propensity for requesting such reports.

Judges were also asked their evaluation of the quality of information available to them at disposition. Sixteen of the 23 judges offered an "excellent and complete" rating; five offered "good but incomplete"; none rated information as "poor."

After initial disposition, juvenile court judges have the option to modify their prescriptions. In fact, the judges' power to modify their dispositional orders is a vital feature of the juvenile justice system. However, exercise of this power requires that judges be aware of subsequent changes of developments in the juvenile's status. Judges were asked about stimuli to initiate reviews of particular juvenile cases. The choices included probation worker reports, petitions from the juvenile or his family, court-appointed inspector or ombudsman reports, personal inspections or investigations (by the judges themselves), routine reports by a juvenile corrections authority or institution,[12] special recommendations by a juvenile corrections authority or institution, reports or petitions of independent community or professional groups, or media reports. Again, probation worker reports were ranked the highest (most consistently in New York). Fifteen responding judges placed this source first, one placed it second, one placed it third, and three checked it as important but gave no ranking. Petitions from the juvenile or his family and reports from a juvenile corrections authority also were important, receiving rankings of second and third from roughly half the respondents. Media reports or court-appointed inspectors were considered irrelevant by all respondents, and personal investigations by judges were attached some importance by eight respondents.

Judges need information not only about the juvenile but also about the quality and effectiveness of alternative suppliers of services. When asked what sources they depended on for this kind of information, judges ranked probation worker assessments highest again, with 12 first-place votes, two second-place votes, and six indications of importance from judges who failed to rank the alternatives.

In Suffolk County while probation is the most important source of information about available programs and institutions, the Probation Department does not undertake any systematic evaluation of alternatives. Since probation workers have frequent contact with the schools and institutions in connection with their investigation and placement work, they rely on this experience to provide impressions of the quality of institutions and programs. In New York State, judges themselves are encouraged by law to visit schools and institutions they make use of for disposition;[13] these visits, arranged by the Probation Department, serve to supplement secondhand information with firsthand impressions.

The Probation Department in Suffolk does not systematically

TABLE 2
SOURCES OF INFORMATION ON
ALTERNATIVE INSTITUTIONS AND PROGRAMS

	Source of Information					
Rankings Given	Probation Worker Evaluations	Complaints From Juveniles or Families	Court's Own Staff Evaluation Studies or Inspections	Personal Inspection by Judge	Routine Reports by Juvenile Corrections Authorities or Institutions	Media Reports
No. of judges ranking source as No. 1	12	1	1	1	2	0
No. of judges ranking source as No. 2	2	3	4	6	4	0
No. of judges ranking source as No. 3	3	2	1	5	4	0

search out and compile information on new alternatives (agencies and programs) that may become available as dispositional options. The New York State Board of Social Welfare is required to approve new agencies. The Court and Probation Department are notified of newly approved facilities, and probation follows up such notices with on-site visits.

In general, other sources than probation, including complaints from juveniles or their families, routine reports by juvenile corrections authorities or institutions, and personal inspections by judges, were also deemed relevant by judges in the mail survey, as illustrated in Table 2.

EXPERTISE

A striking feature of the juvenile court judges' work, in contrast to the work of other kinds of judges, is the strong psychological/social-work character of the decisions they must reach.[14] We have noted the access and relatively high utilization that surveyed judges make of expert opinion (probation workers, psychologists, and psychiatrists) in these fields. Yet, it is not clear that judges have the necessary background to evaluate such information. Of the 23 judges responding to the survey, 21 indicated they had law degrees. (One of the two responding "judges" without a law degree was actually a "juvenile referee," not a full-fledged judge.) Only two judges indicated they had graduate degrees in other fields (in this case education and sociology). This conforms to the findings of the 1965 National Survey which found that only 8% of juvenile court judges had graduate degrees in other fields. There are the incumbent dangers, therefore, that judges may either ignore, or blindly follow, the expert opinions of behavioral science professionals.

Conversations with judges in Suffolk County indicate that their "working" knowledge in the behavioral sciences comes through "on the job training"; judges learn to evaluate individual professionals by their "track records" in providing successful recommendations over time (i.e., judges learn whose recommendations seem to be worth following). These modes of learning by judges may be particularly costly if job turnover is rapid for juvenile court judges. On this matter, surveyed judges ranged in time spent on the juvenile court bench from six months to 33 years, averaging eight years. Not all judges were full-time on the juvenile bench, however; 10 of the judges responding to the survey said they were full-time in the juvenile or family court division. This proportion (45% of surveyed judges) is much higher than the 5% full-time figure found in the 1965 National Survey.

RESOURCES

Judges were asked how much time they spent on individual juvenile cases and whether they felt their caseloads severely limited the time they could devote to individuals. Twelve judges felt that they were severely limited; nine said they were not; two did not respond to this query. Estimates of minimum time spent on any individual juvenile ranged from 10 minutes to one day per year; estimates of maximum time spent on any given individual ranged from one-half of a day to 10 days.

Since the court does not normally pay directly for the treatment services it prescribes at disposition, no questions were posed to judges regarding the budget available to pay for these services. Judges were asked, however, what implications they thought their decisions had in terms of financial impact on treatment suppliers. Judges in

California noted the probation subsidy program, which compensates the county for 50% of the cost of each juvenile placed on probation.[15] Other judges noted the per diem payments received by private schools and institutions that accepted referrals of the court. The financial effects of placement at public institutions are apparently less clear-cut; placement policies may affect costs at these institutions (as one responding judge noted) but there may not be corresponding changes in revenues received.

ALTERNATIVES

Judges were asked what alternatives they had available for placing juveniles. Table 3 summarizes their responses.

Certain options seem widely available, such as probation, training schools (with the notable exception of Massachusetts),[16] foster care, and work camps. Others, such as halfway houses and public group homes and residences, are available on a more limited basis but are on the increase. Judges were asked if they had the option to turn juveniles over to a department of juvenile corrections and to leave the placement decisions to that authority. Seventeen judges responded affirmatively, indicating they used this option for repeated offenders or "hard core" cases where other alternatives have failed. A striking contrast appears in the use of this option between responding judges in California and those in New York. California judges indicated they use the option in 1% to 5% of cases, whereas New York judges estimated usage in 10% to 30% of cases.

Judges were asked whether they actually have used all the alternatives they said were available to them. All (18) judges

TABLE 3

Alternative	Number of Judges Having This Alternative Available
Training schools	20
Probation	23
Private group residences (13 to 25 children)	16
Public group residences (13 to 25 children)	10
Halfway houses	5
Private group homes (4 to 12 children)	17
Public group homes (4 to 12 children)	7
Foster care	23
Forestry or other work camps	19
Special part-time vocational, educational, drug or other rehabilitative programs provided in the community	19

responding to this question said they did. When asked if they were necessarily limited to the options listed in the previous table, 12 judges said "yes" and 11 said "no." When asked if there was some way to cultivate new options and circumvent traditional supply channels, 14 responded affirmatively, indicating such options as fostering local legislation, helping secure federal grants for community facilities, and other local initiatives aided by the court. Thirteen judges indicated they have tried, or are trying, such options.

LEVERAGE

Judges were asked what actions they have taken to stimulate corrective action for institutions and programs they felt were performing poorly. Twenty of the judges indicated they changed their dispositional patterns, i.e., they used alternatives to the programs or agencies they thought were unsatisfactory. (This would be corrective in nature insofar as the treatment institutions were sensitive to losses in patronage.) Fifteen judges said they complained or directed pressure to the juvenile corrections authority. Eight judges indicated they have used the authority of the court to order changes in juvenile treatment programs or institutions. As we mentioned earlier, some judges actively engaged in initiatives to institute new programs; three judges saw this as a mode of corrective action. One judge noted that he made speeches and issued press reports to obtain desired improvements.

CONSTRAINTS

Twenty of the responding judges checked "overcrowding" as a barrier limiting their discretion to place juveniles where they choose. This factor can be considered one of several "economic" constraints stemming from a lack of public resources to purchase adequate care. Several judges mentioned other factors, such as public budget constraints and the high prices of private care combining to limit the use of private agency options, as well as the selectivity of private institutions that can afford to reject court referrals.

Judges also acknowledged limitations to their own legal authority in juvenile corrections decisions, including the right (of the juvenile or his family) to appeal to a higher court (rarely exercised). Furthermore, while 22 of the 23 responding judges stated they had general authority to amend a disposition subsequent to its implementation, several judges noted that this option was limited to reducing the severity of the "sentence." Also, some judges in New York and Massachusetts noted their loss of discretion once placement is made with a state authority or institution. This loss can result in a dismissal of a juvenile from treatment without the court's knowledge.

OBJECTIVES

Responding judges seem fairly evenly divided in how they perceive their judicial versus correctional roles. The characterization "objective interpreter and administrator of law" was most highly ranked, followed closely by "proxy-guardians for juveniles." Judges nevertheless appear to reconcile these roles in terms of objectives seeming to serve both purposes, principally "rehabilitation." Table 4 summarizes the judges' rankings of their role perceptions and objectives.

In terms of modus operandi, one type of activity that would differentiate the judicial from correctional aspects of the juvenile court judges' function is their active participation in "post-disposition" decisions regarding the juvenile. Judges were asked what factual information had led them in the past to modify a juvenile's current dispositional status. Table 5 indicates their responses. It appears that judges *do* see themselves playing an active role in

TABLE 4

Role Characterization or Objective	Number Judges Ranking as No. 1[a]	Number Judges Ranking as No. 2	Number Judges Ranking as No. 3
Roles			
Objective interpreter and administrator of law	12	2	6
Protector of society	1	11	4
Proxy-guardian for juveniles	9	8	3
Objectives			
Punishment	1	0	1
Rehabilitation	21	2	0
General deterrence	2	5	6
Maintaining family integrity	1	10	5
Avoiding institutionalization	1	6	3
Removing dangerous juveniles from community	0	3	6

a. Some judges designated more than one role, or more than one objective, with equal priority (e.g., a judge may have ranked two objectives as No. 1). Also, judges did not rank roles or objectives they considered irrelevant.

TABLE 5

Information Leading to Change in Dispositional Status	Number of Judges Citing This Point
1. Change in juvenile's behavioral pattern, e.g., improved school grades or behavior (failure to get into subsequent trouble)	21
2. Change in juvenile's family situation, e.g., parental break-up or reconciliation; parental acceptance of family counseling	17
3. Indications of maladjustment of juvenile to his prescribed treatment program, e.g., reports that juvenile fails to cooperate with treatment program administration; indications that child requires more structure or supervision, e.g., runaways	16
4. Indication of incompetence or unwillingness of treatment provider to properly serve juveniles' particular needs, e.g., failure to provide counseling; lack of indications of success or achievement; failure of treatment agency to provide information or otherwise cooperate with agency responsible to the court; treatment agency request to expel juvenile it cannot handle	15

correctional decisions, even after the initial court proceedings have been consummated.

ACCOUNTABILITY

By the nature of their appointments, judges are "political" people: they are either elected directly or appointed by elected officials. Table 6 presents a summary of the tenure and nature of appointment of judges responding to our survey.

One can hypothesize that judges elected rather than appointed, with shorter rather than longer tenures, would be more responsive to the electorate, while appointed, longer term tenure judges would be more independent and/or responsive to professional rather than political elements. Judges were asked what parties they are influenced by, or have consulted with, in their approach to juvenile case decisions. Their responses are summarized in Table 7. It is interesting

TABLE 6

Appointment and Tenure	Number of Judges	States[a]
Elected for 6 years	5	California, Illinois
Elected for 10 years	12	New York, Pennsylvania
Appointed for 4 years	1	Illinois
Appointed for indefinite term or life	4	Massachusetts, California

a. Local practices vary in some states; state was unknown for some responding judges, and thus, the number of judges may be different from the number of responses from states listed here.

TABLE 7
SUMMARY OF RANKINGS RECEIVED BY PARTIES TO WHICH JUDGES MAY FEEL RESPONSIBLE[a]

	Nature of Appointment				
	Elected		Appointed		
Responsive to	6 Years	10 Years	4 Years	Indefinite	Total
Fellow judges or masters	1/2	2/5	0/1	3/3	6/11
Other professional peers or associations	1/1	6/11	0/1	1/2	8/15
Juvenile corrections officials	3/3	7/10	0/1	3/4	13/18
Members of the community or community groups	1/3	0/4	0/1	1/2	2/9
Elected officials	0/1	0/0	0/0	0/0	0/1

a. Entries above the diagonal (/) indicate the number of first or second rankings received; entries below the diagonal indicate the number of judges acknowledging this party as a source of consultation or influence. No striking differences between the groups of judges bearing different types of appointment are apparent from the data.

that only one judge (elected for six years) acknowledged elected officials as a source of influence or consultation. In general, juvenile corrections officials (including probation workers) seem to wield the most influence, with fellow judges or other professional peers secondarily important.

V. CONCLUDING DISCUSSION

The results of our small-scale survey and field observations, while by no means conclusive, serve to indicate certain sources of difficulty and areas potentially needing repair in the functioning of the juvenile correctional system.

In terms of the information available to judges, one fact stands out strongly—the judge is highly dependent on the probation service. Probation serves simultaneously as the enforcement arm and "information feedback" mechanism of the court, as well as a source of supervision and treatment. This puts a great deal of responsibility on probation to provide accurate and complete information both with respect to individual juveniles' backgrounds and current progress (even after disposition), and with respect to the quality and effectiveness of alternative agencies available to provide treatment and custodial services. Upgrading of the professional training of probation workers would seem a worthy investment, therefore, where such preparation is lacking. (In Suffolk County, probation workers are required to have bachelor's degrees, pass a civil service test, and take in-service training.)

On the matter of information about individual juveniles, probation "work-ups" seem generally to be well done; if not, judges seem to enjoy wide discretion to secure whatever information or reports they desire before making a disposition. On the other hand, information about the character and performance of alternative sources of treatment services supply is not systematically provided by probation departments or any other agency.

While it is reassuring that most judges surveyed made personal inspections of juvenile facilities, a regular flow of formal evaluation reports, prepared by probation departments or another independent agency, would seem worth considering to keep judges informed about the implications of their choices.[17]

The matter of expertise appears to be a source of concern in juvenile court decision-making. If law degrees are any evidence, juvenile court judges are intelligent people, well versed in the legal aspects of their work. But there seems no cause to be confident in judges' abilities to make good decisions involving social work and psychological treatment. Judges are neither specially trained nor screened for their aptitude or knowledge in these areas. It would seem well to consider proposals for specialized training in law school, in-service training, or other formal educational or experiential prerequisites. Otherwise, as we have previously remarked, there remain clear dangers of uninformed decisions or blind acceptance of recommendations by probation or other professional sources of information or advice. This worry is reinforced by the apparently strong reliance placed by judges on the opinions of correctional officers. The latter poses the specific problem of relying on information from the supplier as to the desirability of the supplier's services.

The dependence of judges on probation workers (and other parties regularly in contact with juveniles) becomes even more apparent when one considers the limited time judges have to spend on any one case. Yet the time available to judges is a scarce resource that cannot be bought in sufficient quantity to allow judges a leisurely consideration of each case. Perhaps a better approach would be to give judges better *control* over the resources they utilize to obtain information and dispositional recommendations.

The question then arises, should probation departments (or the information and investigative parts thereof) necessarily report directly to the (juvenile) court rather than be (part of) an independent bureau in the administration of local government? In Suffolk County, for example, the Director of Probation reports directly to the County Executive. Yet judges seem satisfied that probation is responsive to them in providing whatever information they want. There are several explanations for this. First, its charter defines probation functionally as a service to the court, and traditionally this is how probation has developed. But perhaps more fundamentally, the power relationships are such that, if necessary, judges could discipline the Probation Department by virtue of their influence with county administration. Essentially, probation operates independently only as a convenience to obviate the need for judges to assume the burden of Probation Department administration. But in terms of effective accountability, probation reports to the court; if this ever ceased to be the case, consideration ought to be given to formalizing court authority in this area, to insure judges the informational access they require.

An area where judicial control over resources is a more controversial and basic issue is in the funding of treatment agencies. When judges prescribe services to juveniles they are implicitly spending from constrained public budgets not under their own direct control. When private schools or other agencies are used, explicit per diem payments are made from local government funds; when referrals are made to public institutions, there are definite cost implications and long-run if not short-run effects on public expenditures. The key questions are, thus: (a) are budgeted resources for treatment alternatives large enough and (b) do judges have sufficient control to ensure that these resources are spent effectively? There are indications that both these questions should evoke negative responses.

The facts of overcrowding of public institutions and high selectivity of private schools, cited by judges in our study, serve to indicate possible severe limitations on the funds available to purchase the kinds of services judges would like to prescribe. On the other hand, for the public alternatives at least, it is not clear that existing resources are spent in a manner corresponding to what judges would like to have available. One might consider therefore, proposals to completely "voucherize" the system of funding treatment services for juvenile court referrals. This would essentially entail providing the court with a total budget for juvenile treatment services and allowing the court to spend this budget on a case by case basis as it wished. That is, the court could select the type of service and source of its supply for each juvenile, as long as it kept within its total budget constraint. The court does this implicitly now, for private agency services, but the proposal would broaden this mechanism to include all services and make it explicit. To ensure some regularity, general guidelines might be imposed on "price ranges" acceptable to the public, for judicial purchase of various kinds of custodial and therapeutic services for juveniles.

The concept of "consumer" purchasing of public services with voucher certificates, while novel, is not unprecedented. Food stamps, for example, have become commonplace, and even in the area of public education, experiments are proceeding in the use of this concept.[18] The advantage is to force suppliers to respond to the desires of the consuming public by tying financing to consumer purchases. In the present case, the "consumer" would of course be represented by the court.

The present survey provides some evidence that judges identify with their correctional as well as judicial role, but there is little assurance that judges are accountable on this score. While individual judicial decisions are occasionally reviewed for purposes of updating or appeal, there is no systematic review process that examines the decision-making patterns (and their performance implications) of judges themselves. For, unlike consumers in the private marketplace, judges are spending public money and affecting second parties directly, and, also unlike private consumers, there is virtue in having the decisions of judges fairly *consistent* in terms of the types of treatment afforded juveniles in various age, offense record, sex, race, and other categories.

Several alternatives for review are possible. The simplest would be a peer process in which judges within a court systematically review their cases with one another on a regular formal basis. A more substantive alternative would be to have an independent evaluation agency, possibly under bar association auspices but separate from the court or service suppliers, carry out regular follow-up studies on judicial decision-making and make this information available to judges and (in a form which leaves juveniles anonymous) to the public at large. More extensive procedures for review are also possible of course; one must be wary, however, of options that could severely constrain the discretion of judges to act on individual cases. It is the emphasis on overall, long-term performance of each judge that is of concern in attempting to effect accountability.

The method of judicial advancement does little to encourage

confidence in the accountability of judges. Advancement is through the political reward system, which correlates tenuously with performance on the juvenile bench. Fortunately, professional prerequisites are often attached to eligibility for election or reappointment. In New York for example, a law degree plus a 10-year membership in the bar is required. But no social work or behavioral science requirements exist. For the interested reader there is substantial literature on the selection and advancement of judges.[19]

In summary, there are enough apparent imperfections in the present mechanisms of judicial decision-making of the juvenile court to warrant further investigation and analysis before the desirability of the juvenile-court proxy-consumer model of juvenile correctional arrangements can be generally recommended. Furthermore, other key parts of the juvenile system in addition to the juvenile court–principally probation and the array of public and private supply institutions and agencies–must be analyzed from the viewpoint of political, economic, and bureaucratic behavior. Nevertheless, such a study would appear worthwhile, not only for improving the performance of juvenile corrections but also because the organizational framework of this system may ultimately have some transfer potential to the more monopolistic and repressive domain of adult criminal corrections and to other public service fields.

NOTES

1. See, e.g., In Re Winship (1970); In Re Gault (1967); and Kent v. United States (1966).

2. For background on the history of the juvenile court see Lerman (1970).

3. For example, Maryland placed its Department of Juvenile Services under the aegis of the State Department of Health and Mental Hygiene and set up an advisory board representative of welfare and criminal justice agencies and the general public in an attempt to stimulate coordination and innovation in juvenile services.

4. See Hirschman (1970). "Exit" and "voice" are two modes of reacting to decline in the performance of organizations. According to Hirschman, voice can be defined as "any attempt to change, rather than to escape from an objectionable state of affairs, whether through individual or collective petitions to the management directly in charge, through appeal to a higher authority with the intention of forcing a change in management, or through various types of actions and protests, including those that are meant to mobilize public opinion." In contrast, "exit" is when "the customer who, dissatisfied with the product of one firm shifts to that of another, uses the market to defend his welfare or to improve his position; and he also sets in motion market forces which may induce recovery on the part of the firm that has declined in comparative performance."

5. For background see New York State Special Commission on Attica (1972) and Vanden Heuvel (1972).

6. That the present juvenile system can be repressive is documented by Morales v. Turman (1973) and In Re Gault (1967). For general documentation of inadequacies see the President's Commission for Law Enforcement and the Administration of Justice (1967).

7. For example, the courts, in right-to-treatment suits, have established minimum conditions for supply. See, e.g., Nelson v. Heyne (1972); Marterella v. Kelley (1972); and Morales v. Turman (1973).

8. See Street, Vinter, and Perrow (1966) and Committee on Mental Health Services Inside and Outside the Family Court in the City of New York (1972).

9. See, for example, Skolnick (1967).

10. See, for example, Hogarth (1971).

11. In California, the probation officer has the responsibility of providing a minor's social profile to the court; the court may also request a psychiatric evaluation. Illinois requires that a written report of social history be completed and considered within 60 days prior to any order of commitment. In Massachusetts, the judge may, before disposition, refer a child to the Department of Youth Services, which makes a diagnostic study of the child on an outpatient basis and submits a report and recommendations to the court. Pennsylvania provides for a social and psychiatric study pursuant to a court request. New York requires a social history report to be prepared by the Probation Department, and psychiatric and psychological evaluations are made upon the court's request.

12. In Massachusetts, the Department of Youth Services has the responsibility to make periodic reexaminations of all persons within its control.

13. New York State statutes provide for the payment of expenses incurred in such visits, to be paid by the (local) authorities responsible for paying the judges' salaries. California requires yearly inspection by judges of all facilities.

14. See Ketcham (1961).

15. See also Smith (1971).

16. The Massachusetts Department of Youth Services, under Jerome Miller, closed down all of the state's training schools with the aim of converting to a completely community-based juvenile correctional system. See Massachusetts Department of Youth Services (1972).

17. Studies such as that made by Koshel (1973) serve the purpose of delineating the general effectiveness of alternative modes of treatment, while evaluation reports on specific agencies such as those now being made available by the New York State Board of Social Welfare point out the nature of alternative sources of supply.

18. See, for example, Doyle (1973).

19. See, for example, Winters (1966).

CASES

In Re Gault (1967) 387 U.S. 1.
In Re Winship (1970) 397 U.S. 358.
Kent v. United States (1966) 383 U.S. 541.
Marterella v. Kelley (1972) 349 F. Supp. 575 (S.D.N.Y.).
Morales v. Turman (1973) Civ. Act No. 1948 (E.D. Tex. Order of August 31).
Nelson v. Heyne (1972) 355 F. Supp. 451 (N.D. Ind.).

REFERENCES

Committee on Mental Health Services Inside and Outside the Family Court in the City of New York (1972) "Juvenile justice confounded: pretensions and realities of treatment services." Paramus, N.J.: National Council on Crime and Delinquency.

DOYLE, D. (1973) "The San Jose educational voucher experiment." Washington Operations Research Council Conference on Urban Growth and Development, Washington, D.C. (April).
HIRSCHMAN, A. O. (1970) Exit, Voice and Loyalty. Cambridge, Mass.: Harvard University Press.
HOGARTH, J. (1971) Sentencing as a Human Process. Toronto: University of Toronto Press.
KETCHAM, O. (1961) "The unfulfilled promise of the juvenile court." Crime and Delinquency (April).
KOSHEL, J. (1973) "Deinstitutionalization–delinquent children." Urban Institute Paper 963-15 (December). Washington, D.C.: Urban Institute.
LERMAN, P. [ed.] (1970) Delinquency and Social Policy. New York: Praeger.
Massachusetts Department of Youth Services (1972). A Strategy for Youth in Trouble. Springfield, Mass.
McCUNE, S. D. and D. L. SKOLER (1965) "Juvenile court judges in the United States, Part I: A national profile." Crime and Delinquency (April).
National Council of Juvenile Court Judges (1972-1973) Juvenile Court Judges Directory. Reno: National Juvenile Court Foundation.
New York State Special Commission on Attica (1972) Attica: The Official Report of the New York State Special Commission on Attica. Albany, N.Y.
President's Commission for Law Enforcement and the Administration of Justice (1967) Task Force Report: Juvenile Delinquency and Youth Crime. Washington, D.C.: Government Printing Office.
SKOLNICK, J. (1967) Justice Without Trial: Law Enforcement in Democratic Society. New York: John Wiley.
SMITH, R. L. (1971) A Quiet Revolution: Probation Subsidy. California Youth Authority.
STREET, D., R. D. VINTNER, and C. PERROW (1966) Organization for Treatment. New York: Free Press.
VANDEN HEUVEL, W. (1972) "The press and the prisons." Columbia Journalism Review (May/June).
WINTERS, G. R. (1966) "Selection of judges–an historical introduction." Texas Law Review (Summer).

3

Fiscal Equalization Through Court Decisions: Policy-Making Without Evidence

ROBERT L. BISH

□ A LONG TRADITION in local government finance literature decries the fiscal disparities and mismatch between needs and resources among local governments in metropolitan areas. During the past few years debates over the issue of fiscal equality have moved into the courts. Significant and far-reaching decisions have been made, primarily with regard to school finance. These court rulings contain both positive and normative analyses: positive analysis in that they postulate certain causal relationships among variables such as property values, tax rates, and fiscal effort, and normative analysis because they state that certain criteria of equality must be met because education is of fundamental importance.

This chapter will bring social science evidence to bear on the postulated relationships and predictions inherent in proposals to provide equal opportunity through the fiscal reform that is both advocated by lawyers bringing the school finance cases and designated as meeting equal opportunity criteria by judges in their rulings. This examination includes some important observations on (a) the relationship between income redistribution and school finance

AUTHOR'S NOTE: *I want to thank William R. Andersen, Frances P. Bish, Julie Johnston, Frank I. Michaelman, Hugh O. Nourse, and Elinor Ostrom for their comments on an earlier draft. Their efforts have led to significant changes in my initial analysis.*

reform and (b) the general problem of income redistribution via equalizing grants for local government programs in metropolitan areas. In the course of this analysis it will become apparent that school fiscal reform efforts are not as related to traditional fiscal reform efforts as they first appear.

FISCAL INEQUALITY AND SCHOOL FINANCE

FISCAL INEQUALITY

Different local government units in metropolitan areas contain different kinds and amounts of economic resources. Some units contain concentrations of highly valued industrial and commercial property; other units are primarily residential. Some units have residents with relatively high personal incomes; others have residents with very low personal incomes. Different measures of economic resources–value of property, personal income of residents, or business income–make different units appear relatively richer or poorer, but all measures indicate that local governments are vastly unequal in their economic resource bases (Advisory Commission on Intergovernmental Relations [ACIR], 1973, 1971, 1967, 1965; Riew, 1970).

Fiscal differences among political units in metropolitan areas are considered by many to be responsible for critical problems (Ecker-Racz, 1970; Maxwell, 1969; ACIR, 1973, 1971, 1967, 1965). One of these problems is the separation of resources from "needs." Jurisdictions that have special expenditure requirements (i.e., cities with large welfare populations or school districts with many disadvantaged students) often lack the resources to overcome their special problems. Another problem is that different political units have unequal tax rates, and unequal tax rates are viewed both as unfair and as distorting the location of business and households from what would otherwise be more efficient location patterns. A third problem is that tax competition to attract business or households forces local governments to keep taxes, and hence public spending, lower than what would otherwise be desired–thus contributing to the general problem of revenue insufficiency in the face of rising costs for all local governments. Also related to tax competition is the allegation that suburbanites "exploit" the central city because they may work, shop, or use civic facilities in the central city but pay taxes to their suburban jurisdictions.[1]

Proposed remedies to these problems have focused primarily on consolidations of small political units, increased state or federal government financing of some functions (such as welfare and education), and equalizing intergovernmental grants. Some movement toward these remedies exists in all states, but even with the programs that have been developed considerable fiscal inequality remains.

SCHOOL FINANCE

Public school finance typifies local government fiscal problems associated with fiscal inequality. In addition to problems of inequality, however, rapid increases in expenditures have added to school finance problems. During the past 15 years expenditures have more than tripled—with three-fourths of the increase due to cost increases and only one-fourth due to growing enrollments. During the past few years costs have continued to rise while enrollments have fallen, and taxpayer approval of school bond and operational levies has fallen—from over 70% success in the middle 1960s to less than half by 1971 (Reischauer and Hartman, 1973).

While public education is primarily a local government function, in 1971-1972, 39% of all revenues were provided by state governments and 6% by the federal government. Local governments provided the other 55%, 82% of which was raised from property taxation. In terms of rapid cost increases, voter resistance to higher taxes, and reliance on property taxation for local government's share of financing, education is like most other locally provided goods and services. However, education does receive much more state and federal support than do other functions, except welfare and highways.

State aid for education follows three patterns. First, there are grants for special programs, such as vocational education or aid for blind and deaf children. Second, there are equalizing grants, usually based on the property tax wealth of school districts, with lower wealth districts receiving greater aid. Third, there are flat per-student grants for all school districts.

All federal aid to local school districts is through functional grants, a major one of which is Title I aid to school districts with students from low-income families. While Title I may provide as much as $300 per low-income student, Title I funds have often simply been used to displace local funds rather than to provide for expanded opportunities for low-income students, and appropriations often have been insufficient to extend aid to all eligible students (Hughes and Hughes, 1972).

While state and federal aid provides some equalization among school districts, educational finance is still determined independently in each state. Thus some states provide nearly all support for education, e.g., Hawaii with 89.4%. Others provide less, e.g., New Hampshire with only 9.9%. Federal contributions also vary—from 28.1% of total expenditures on education in Mississippi to 3.5% in Wisconsin. Because distribution formulas within states also vary, different districts within states also receive greater or lesser shares of state aid. The result is that there is considerable variation in tax rates and expenditures among school districts within a single state and among districts in different states. Variances in tax rates and expenditures by factors of three or four (in different directions) by different school districts within a single state are not uncommon (Oldman and Schoettle, 1974: 941).

Although it has been traditional for persons concerned with educational finance to decry these differences, especially when low-wealth districts with high tax rates raise considerably less money than wealthier districts with lower tax rates, it was only with a series of court decisions beginning in 1971 that real "reform" in educational finance appeared likely. An examination of the rationale underlying these court decisions and proposed remedies provides considerable insight into traditional analyses of local government fiscal inequality, especially the neglected importance of land rent theory.

LEGAL CHALLENGES TO SCHOOL FINANCE

Unequal capacities among school districts was tested in a 1971 California court case, Serrano v. Priest. In this case the California Supreme Court ruled that expenditure on education could not be a function of the wealth of a school district because of the Fourteenth Amendment to the U.S. Constitution and similar provisions of the California State Constitution.[2]

The evolution of the Fourteenth Amendment to apply to school financing in Serrano v. Priest (1971) involves several important legal concepts. These concepts are those of "fundamental interest," "suspect classification," "strict scrutiny," "compelling purpose," and "rational purpose."[3] "Fundamental interests" are interests that are either specifically guaranteed by the Constitution or declared by the Supreme Court to be of such importance to individuals or value to society as to assume such stature. Fundamental interests, defined by the Court in relation to the Fourteenth Amendment, include the right to an equal weight vote (Reynolds v. Sims, 1964), the right to appeal criminal processes regardless of wealth (Griffin v. Illinois, 1956), the right to interstate travel (Shapiro v. Thompson, 1969), the right to procreation (Skinner v. Oklahoma ex. rel. Williamson, 1942), the right to marital privacy (Griswold v. Connecticut, 1965), and the right to court access for divorce proceedings (Boddie v. Connecticut, 1971).

A "suspect classification" is a classification of individuals which serves no useful purpose and which often implies a stigma or majoritarian abuse toward the group so classified. The classic suspect classifications are race (Strauder v. West Virginia, 1880) and national ancestry (Korematsu v. U.S., 1944).

"Strict scrutiny" and "compelling purpose" are directly related to fundamental interests or suspect classifications. Whenever a government program or policy infringes upon a fundamental interest or results in a suspect class being subject to a different treatment, the constitutionality of that program or policy is subject to "strict scrutiny" by the court. To meet the test of strict scrutiny the program or policy must serve a "compelling purpose," which means that the program or policy is *necessary* to the achievement of an extraordinarily important public objective. If alternative ways can be identified to achieve the public objective, the program or policy which infringes on a fundamental interest or involves differential treatment of a suspect class is unconstitutional. If, however, government programs or policies neither infringe upon a fundamental interest nor differentially treat a suspect class, the program or policy has only to serve a "rational purpose," that is, be related in some reasonable way to the achievement of a public goal.

Education, in Serrano v. Priest (1971), was drawn under the Fourteenth Amendment in two ways. First, the California Supreme Court held that education was of such importance in America that it deserved the status of a fundamental interest under both the U.S. Constitution and the California Constitution; second, it declared that wealth of a school district was a suspect classification. Both of these determinations, either of which would force the court to apply strict scrutiny to determine the constitutionality of the system of public school finance in California, were precedent setting. This was the first declaration of education as a fundamental interest—although the importance of education had been stressed in previous decisions, such as Brown v. Board of Education (1954). Designation of wealth as a basis for suspect classification was not totally new because the lack of wealth preventing an individual from enjoying a fundamental interest is present in several of the fundamental interest cases including Griffin v. Illinois (1956), where it was ruled that court transcripts for appeals had to be provided free to persons who could not afford to pay for them, and Shapiro v. Thompson (1969), where it was ruled that the one-year residency requirements for public welfare were unconstitutional because they prevented interstate travel. The big break with previous cases, however, was that previous cases had identified the poverty of a *particular* individual or family in relation to a fundamental interest as a suspect class. Nowhere had wealth stood alone to compel strict scrutiny of a government program, and never had the concept of suspect classification been applied to the heterogeneous residents of a government jurisdiction, which contained within its boundaries low levels of wealth in terms of property subject to taxation. The Serrano court assumed that there was a correlation between school district wealth and the income of residents—the issue was not explored nor was evidence brought to bear on this question.

Given its designation of education as a fundamental interest and the wealth of a school district as a suspect classification, the California Supreme Court ruled that if, under the current system of public school finance in California, the low level of wealth in a school district affected expenditures for education, the system was unconstitutional unless it served a compelling purpose, i.e., there was no other way to finance public schools in California. The Supreme Court then remanded the case back to the Los Angeles Superior Court for a trial to determine "facts" and search for remedies.

Before the Superior Court trial of the Serrano case, however, a similar case from Texas reached the U.S. Supreme Court (Rodriguez v. San Antonio Independent School District, 1972). The U.S. Supreme Court did not follow the line of reasoning in Serrano on either critical point. First, it ruled that it was not the court's job to define new "fundamental interests," no matter how important the justices thought them to be. Second, it ruled that a heterogeneous group of people in a school district characterized by low assessed valuation of property within it was not a suspect class. In reaching this latter decision, it specifically examined empirical evidence on the relationship between the income of people and the wealth of school districts and concluded both that poor people did not necessarily reside in poor school districts (or rich people reside in wealthy districts) and that there was no precedent or rationale for declaring the heterogeneous population of a low-wealth school district, in contrast to people characterized by low income, as a suspect class. After the court rejected education as a fundamental interest and wealth of school districts as a suspect classification, it found that the public school finance system in Texas met a rational purpose in spite of fiscal and expenditure inequalities (San Antonio Independent School District v. Rodriguez, 1972).

The Rodriguez decision was made prior to the remanded Serrano trial, and thus was taken into account by Judge Bernard S. Jefferson in his "Memorandum Opinion Re Intended Decision" of April 10, 1974. This memorandum is in effect a decision that may be appealed, but it is not a final decision in that the court has retained jurisdiction of the case until the legislature has a reasonable time to enact a new public school finance system that meets the tests of constitutionality. This memorandum shifts emphasis from the earlier Serrano decisions in two areas. First, in light of Rodriguez, it was determined that education could be declared a fundamental interest under the California State Constitution—even if not under the U.S. Constitution. Second, on the basis of a few school district observations selected by the plaintiffs, the court accepted the idea that poor people resided in poor districts and that people in poor school districts could be considered a suspect classification in relation to education in California. The court then concluded (1) that different levels of education are provided for children in different school districts and (2) that the different educational levels are related to the wealth of the school districts, with persons residing in low-wealth districts having to pay higher tax rates to achieve education at an expenditure level lower than that achieved by persons residing in wealthier school districts who paid lower tax rates. Thus, a suspect classification (wealth) prevented achievement of a fundamental interest (education) and the current public school finance system had to be held unconstitutional. The judge also reaffirmed a decision by the original trial court—that the case was justiciable, that is, a court-imposed remedy was appropriate and feasible. In reviewing the feasibility of remedies, however, he determined that it was not possible to make decisions on the quality of education provided to students by school districts because measurement problems were too severe—a problem that had also appeared in a previous case (McInnis v. Shapiro, 1968). The judge did decide, however, that the quality of education received by students was closely enough related to educational expenditures (although the expert testimony during the trial was conflicting on this point) so that expenditure per student could serve as a measure of the level of education provided. The judge then indicated four potential reforms that would satisfy the strict scrutiny criteria by removing the influence of the school district wealth expenditures for education provided to children in that district, but he left to the legislature the task of designing and implementing an appropriate system.

The four reforms specifically mentioned include (1) statewide financing of all educational expenditures; (2) consolidation of school districts until each district has a level of taxable wealth equal to the statewide average; (3) separation of commercial and industrial property from the residential property tax base, with revenues from commercial and industrial property collected and dispersed on a statewide basis; and (4) district power equalizing, a system whereby the same tax rate would raise the same number of dollars per student in every school district regardless of that district's wealth level. Of these four mentioned remedies, only power equalizing is relatively new; it was developed by lawyers involved directly in the school finance cases (Coons et al., 1970).[4] District power equalizing has been the most attractive remedy both to advocates and to opponents in the school finance cases because it permits continued local control of expenditures but supposedly eliminates the influence of "wealth" in school expenditure decision processes.[5] None of the parties involved in the cases appears to favor either full statewide decision-making and financing of school districts or extensive consolidation of existing districts. Separation of business property from the local school district tax base for use by the state to equalize fiscal resources among school districts seems to be considered inferior to power equalizing, statewide financing, or consolidations.

Before turning to an analysis of evidence related to the court diagnosis and prescriptions to resolve the problem of fiscal inequality among school districts, it should be noted that the issues in school finance, thus far, parallel the traditional diagnosis of problems associated with fiscal inequality among local governments. First, poor districts were considered unable to finance adequate levels of education while wealthy districts could finance education with a minimal tax effort; thus, there existed a mismatch between resources and needs. Second, different school districts had different tax rates to finance similar levels of education—a fact which was considered "unfair."[6] Because of the parallel diagnoses between school finance problems and traditional local government fiscal inequality problems, a careful analysis of the evidence—upon which the problem diagnosis was made and the prescriptions proposed for school finance—allows considerable insight into the potential for broader local government fiscal reform.

ISSUES AND EVIDENCE

Five issues appear critical to the diagnosis and prescriptions in court-directed school finance reform. These are issues surrounding (1) defining education as a fundamental interest; (2) defining a suspect class by school district wealth; (3) postulating causal relationships among school district wealth, tax rates, and expenditure on education; (4) identification of the relationship between educational expenditure and "outcomes" from educational processes; and (5) the justiciability of the case, that is, ability of the court to designate remedies for plaintiffs.

The issues surrounding education as a fundamental interest and the relationship between educational expenditure and "outcomes" are considered extensively in court cases and law journals (Serrano v. Priest, 1974; Harrison, 1974; Note, 1972, 1969; Schwartz, 1973-1974). Because of the U.S. Supreme Court ruling in Rodriguez, the determination of education as a fundamental interest depends on separate interpretations of the 50 separate state constitutions. The Serrano ruling that education provided is directly related to educational expenditure is not supported by considerable analysis (Averch et al., 1972), but the issue received explicit consideration and testimony from social scientists who should be knowledgeable on empirical evidence surrounding that issue. Justiciability, at least in California, has depended on the determination that educational expenditures are closely enough related to outcomes to use expenditure data in the analysis and evaluation of prescriptions instead seeking "outcome" data, which may not be adequate enough to make the case justiciable. Following a more detailed examination of the remaining issues—defining a suspect class and the postulated relationships among wealth, tax rates, and expenditure on education

—I will turn to an analysis of district power equalizing, the reform which appears most attractive to school finance reform advocates.

SCHOOL DISTRICT WEALTH AND SUSPECT CLASS

Plaintiffs in school finance cases have related school district wealth to suspect class in two different ways. First, some have argued that low-wealth school districts are comprised of low-income residents (Rodriguez v. San Antonio Independent School District, 1972). Second, some plaintiffs have argued that any person residing in a low-wealth school district is a member of a suspect class because he or she is forced to pay higher tax rates to achieve comparable education to that obtainable with lower tax rates by persons in districts with higher assessed valuation (Serrano v. Priest, 1971, 1974). Even plaintiffs emphasizing the second position, however, have selected school districts for adjudication wherein low-wealth districts were populated by low-income persons, and in Serrano v. Priest (1971) the second position is argued by the plaintiffs while the judge's ruling indicates he accepts the first proposition as well.

The relationship between family income and school district wealth is not understood by many analysts of school finance. Even though there is a positive correlation between family income and the value of residence, approximately one-half of the property tax base is comprised of industrial, commercial, agricultural, or other nonresidential property. Thus, if a person resides in a district that also contains much industrial or business property, the wealth of that school district is likely to be largely determined by the industrial and commercial property—even if the person is poor in both income and property. On the other hand, if a person resides in a district that is exclusively residential, even if the residents are wealthy in both property and income, the wealth of the district is likely to be lower than that found in districts with significant nonresidential components in their property tax base. Thus, the only way to identify the relationship between school district wealth and resident's income is with empirical research. It is in the bringing to bear of such empirical evidence that analyses in law journals and court cases have, for the most part, been seriously deficient.

The reasons underlying the deficient analyses are easy to identify. While data on assessed value for school districts are readily available, data on income of residents by school district are not. The smallest area for which data are easily available on family income is the county. Thus, analysts take wealth and income data by county area and conclude that wealth and income are correlated (Harrison, 1974). This conclusion for counties is correct, as are similar analyses at the state level. However, as the geographic scale of the unit under analysis becomes smaller, the land uses within the units become more specialized. Divergences between wealth as measured by total taxable property and family income become more predominant. Unfortunately, most analysts of school finance either have not had the resources to determine family income by school district or have not recognized the importance of undertaking their analyses at the school district level instead of at the county or state level. In California, a Ford Foundation funded study did provide sufficient resources to relate school district wealth with family income within school districts. The correlation identified was less than 0.02 for any kind (elementary, secondary, or unified) of school district (Cox, 1975).[7]

Similar observations have been made for Connecticut (Note, 1972), Texas (San Antonio Independent School District v. Rodriguez, 1972), and Washington (Andersen, 1975). Separate analysis is needed in each state because the sizes of school districts in different states vary. For example, in the South where school districts are the size of counties, one would expect a higher correlation between income and district wealth than in states where school districts are smaller. It is clear, however, that *one cannot conclude that low-wealth school districts are populated only or even primarily by poor people.* Thus to define the people residing in a low-wealth school district as a suspect class depends on postulated relationships between wealth, tax rates, and expenditures—not on the income of the residents.

WEALTH, TAX RATES, AND EXPENDITURES

Analyses of the relationships among wealth, tax rates, and expenditures that have been presented in school finance cases have been demonstrations of simple correlations between wealth and tax rates and between wealth and expenditures. The causal relationship postulated is that high wealth results in low tax rates and high expenditures, and conversely, that low wealth results in high tax rates and low expenditures. This assertion of causality implies that property value (wealth) is determined independently of property taxes or educational expenditures. That is, property values do not change in response to changes in either taxes or expenditures, but taxes and expenditures change in relation to changes in wealth. This presupposition is critical to some of the important conclusions in educational finance. This presupposition, however, is neither warranted by the major body of theory which treats property value determination—land rent theory—nor supported by empirical evidence on changes in property value in response to changes in property taxes or educational expenditures. The direction of causality suggested, both by land rent theory and by empirical evidence on the relationships among taxes, expenditures, and property values, is that both property taxes and expenditures are partial determinants of property values and that changes in taxes or expenditures will result in changes in wealth as measured by property value. These relationships among property values, taxes, and expenditures are important for understanding the impact of both school finance reform and local government fiscal reform.

LAND RENT THEORY

The basic tenet of land rent theory is that property values are determined by the benefits and costs associated with the use of a piece of property (Bish and Nourse, 1975: chapter 4). Both current and future benefits and costs are taken into account. Benefits and costs are not only related to the physical characteristics of the property, but to *anything* that makes the property more or less desirable. Property with a central location, for example, is worth more than property less centrally located; property with good public services is worth more than property without good services; and property with low taxes is worth more than property with high taxes, other things being equal. It is even possible for the costs associated with a piece of property to exceed the benefits so that the property is worth less than nothing—which results in abandoned property, as is found in many large cities.

Property markets are extremely complex, and the value of any piece of property is determined by many factors. Our concern here is specifically with property taxes, educational expenditures, and, most important, the impact of school finance reform as represented by district power equalizing; thus empirical evidence focusing on taxation and education expenditures will be examined here. However, studies that relate other factors to property values—factors such as travel time to employment centers, cleanliness of the air, socioeconomic characteristics of neighborhoods, and proximity to neighborhood parks—could also be used to demonstrate the general warrantability of land rent theory.

Property Taxes and Property Value. Property taxes affect the value of property in two ways. First, short-run changes in taxes are reflected in a direct offset in the value of the property. Second, over the long run some of the short-run effects will be partially offset by adjustments in the supply of property improvements.

The short-run change in property value resulting from a change in costs or benefits associated with a piece of property is called capitalization. A one-time increase in property taxes, which is expected to be permanent and is unaccompanied by any other change, for example, will be capitalized into the value of the property. Such a tax increase will result in a lower property value because any buyer would view the total cost of the property as including a larger tax payment. The larger tax payment must be offset by a lower price if benefits from use of the property are to remain unchanged. Property tax capitalization is easiest to observe on apartment buildings where net income is directly affected by a change in taxes; the market value of the building is actually calculated directly by capitalizing its net income stream.[8]

In the long run an increase in property taxes will reduce the amount of new construction, and the reduced supply of buildings will result in higher net income and thus higher values for existing buildings—partially offsetting the impact of the original tax increase. If a public policy is directed toward either raising or lowering property taxes everywhere, long-run effects generated by increases or decreases in new construction can be viewed as important enough to perhaps outweigh short-run capitalization effects for policy purposes. However, if taxes are to be raised in some areas and lowered in others, the long-run effects will consist of a shift in new construction from tax-increase areas to tax-decrease areas with net effects on prices to residents washing out or, at minimum, requiring expensive detailed analysis to identify. In this case—as is the case with district power equalizing or other fiscal reforms that raise taxes in some jurisdictions and lower taxes in others without drastically lowering or raising total property tax revenues—the short-run effect of capitalization of tax changes into property values will be dominant.

Several recent empirical studies directly verify the capitalization of changes in property taxes into property values for single family dwellings. Wicks et al. (1968) analyzed changes in property values in response to a property tax rate change on residential property in Missoula, Montana. Their conclusions were that the change in property taxes was directly offset by changes in real estate values and that the change was in the value of the houses as well as of the land. Smith (1970) reached identical conclusions in his study of a property tax change on single family housing prices in San Francisco. In still another, similarly designed study of a small California city, Church (1974) reached the same conclusions–that property tax changes were capitalized into housing values. These studies bring the best evidence to bear on capitalization because they focus directly on the selling prices of particular houses before and after a property tax change.

Other attempts to identify capitalization of property taxes into property values have relied on cross-sectional approaches whereby, using multivariate analysis, an attempt is made to isolate the effects of property tax differences on housing value or rent. It is difficult to draw conclusions from cross-sectional studies because results depend both on the accuracy of data that are known to be imperfect and on the statistical techniques used, and causality cannot be inferred.

In one such study, Orr (1968) examined the effects of different property tax rates on rents in the Boston area. If taxes were directly passed on to renters and not capitalized into the value of the building, rents should have been higher, other things equal, in areas with higher taxes. If taxes were capitalized into the value of the building, rents should not have varied systematically with property taxes. His conclusion was that different property tax rates on different properties in different local government jurisdictions did not significantly affect rent prices. Orr's techniques have been debated, and his conclusions have been questioned (Heinberg and Oates, 1970) and defended (Orr, 1970). Critics do not argue that no capitalization is identified–only that Orr may or may not have identified it. However, in an extension of the Orr study to the city of Boston, Black (1974) also concludes that rents do not vary with property taxes, a result which supports Orr's conclusions.

In other studies utilizing cross-sectional techniques Pasour (1973) identifies capitalization of tax differences in per-acre farm values in North Carolina. Also in North Carolina, Hyman and Pasour (1973b, 1973a) failed to find capitalization of property tax differences between cities for either single-family houses or apartments. While Hyman and Pasour's results have been criticized (Cowing, 1974) and defended (Hyman and Pasour, 1974), their failure to identify capitalization appears to result from the difficulty of identifying a very small variation in property taxes among the different cities in cross-sectional analysis.

In conclusion, evidence is strong and unequivocal that capitalization of property taxes into property values occurs in response to a change in property taxes. These are also the studies with the fewest statistical problems and the ones most relevant for analyzing the impact of the changes in property taxes that would accompany school finance reform with proposals such as power equalizing. Evidence from cross-sectional studies is mixed, but one cannot conclude that there is no capitalization, only that the difficulties of cross-sectional analysis make it difficult to identify. In general, one must conclude that the weight of the evidence on capitalization is that property taxes are a partial determinant of property value and that changes in property taxes will be capitalized into property values—thus verifying the general warrantability of land rent theory.

Education Expenditures and Property Values. Land rent theory predicts that property values will change in response to a change in benefits associated with property, as well as with costs. It has long been accepted that physical improvements, such as sidewalks, street lights, sewers, or paved streets result in increased value to adjacent property, but it is only recently that empirical analysis of the relationship between human services, e.g., education, and property values has been undertaken. In a study of school districts in New Jersey, Oates (1969) was able to identify systematic variations in residential property values associated with educational expenditures. Housing values were higher in school districts where educational expenditures were higher and lower in school districts where educational expenditures were lower. Oates's analysis, however, like cross-sectional studies of tax differences, is not as precise as the studies of tax changes and is further complicated because educational expenditure is not a precise measure of the benefits parents may feel thay obtain from sending their children to schools in a particular district.

In another study Sabella (1974) compared changes in property taxes and changes in expenditures for education with changes in single-family home values. His conclusions were that increased expenditures for education consistently result in higher property values and that increased taxes consistently result in lower property values. His results, like the other studies of tax and expenditure capitalization, lend strong empirical support to the general warrantability of land rent theory.

The implications from land rent theory and supporting empirical evidence are that costs and benefits associated with property are partial determinants of the value of that property. Property value is not determined independently of taxes or benefits as is assumed throughout court rulings and in most of the proposed remedies for school finance reform.

SCHOOL FINANCE REFORM THROUGH POWER EQUALIZING[9]

District power equalizing is proposed to eliminate the effect of wealth on expenditures for education by making the tax price of education the same for everyone (Coons et al., 1970). This means that any given tax rate will raise the same amount of money per student in any district regardless of the district wealth level. For example, if the average wealth per student statewide is $50,000, all school districts would receive revenue per student equal to their tax rates times $50,000[10] The deficits for school districts with actual per student wealth levels below $50,000 would be financed by state collection of surpluses from districts with above-average wealth levels. Thus, no matter how rich or poor a district, the same number of dollars per student would be available for expenditure for each percentage of tax rate throughout the state.

District power equalization is the fiscal reform recommended to achieve equal educational opportunity by supporters of the Serrano decision. From their perspective, if the same tax rate raises the same amount of money per student anywhere in the state, *fiscal capacity* has been equalized with regard to educational finance. A closer look at this conclusion, specifically the meaning of "fiscal capacity" and the relevance of land rent theory, is needed to understand just what power equalizing involves.

Let us examine a simple example of two equal-sized school districts. District A is a "rich" district, containing high-income families, businesses, and commercial property. It has high expenditures on education with a low tax rate. District B is a poor district, containing low-income families and no business or commercial property. It has low expenditures on education with a high tax rate. The total property per student, average residential property value, per student educational expenditures, tax rates, and residential tax bills are presented in Table 1.

Now let us assume that we introduce an equalization program so that each district may set its own tax rate and determine its own educational expenditure level but that the state government shares or supplements revenues collected from school property taxes. Thus, the same tax rate yields the same dollar revenue per student for all districts.

Assume in this case that the state government has decided that for each 1% tax the local school district will be entitled to $400 per student. Districts where a 1% rate would result in more than $400 per student would turn the surplus over to the state government. The state government would then supplement district revenues to achieve the $400 per 1% rate where local property taxes would not raise that amount. Let us also assume that both districts in this example initially wish to retain the same expenditure per student—$1,500 for District A and $480 for District B. Under power equalizing, District A would have to have a tax rate of 3.75% to raise $1,500. This is because for each 1% rate it would receive the state-adjusted $400. Residents of District A would find their tax rate increasing from 1.5% to 3.75% to maintain their $1,500 expenditures per student. This would mean that the average residential tax bill would increase from $750 to $1,875. At the same time residents of District B would be able to achieve their desired expenditure of $480 with a tax rate of 1.2% instead of their former rate of 6%. The average residential tax bill would fall from $600 to $120. The figures for the power equalized situation are presented in Table 2.

These changes in tax burdens, however, are not the end result of the introduction of power equalizing. We anticipate a change in the value of the property to offset any changes in taxes, as discussed in our analysis of land rent theory. Calculation of the impact of a

TABLE 1
SCHOOL TAXES AND EXPENDITURES

	District A	District B
Wealth per student	$100,000	$ 8,000
Average residential wealth	50,000	10,000
Tax rate	1.5%	6.0%
Average residential tax bill	$750	$600
Expenditures on education per student	$1,500	$480

NOTE: We are assuming that each district raises all funds for educational expenditures so that the tax rate times wealth per student equals expenditure per student.

TABLE 2
SCHOOL TAXES AND EXPENDITURES WITH POWER EQUALIZATION

	District A	District B
Expenditure per student	$1,500	$480
Tax rate required at $400 per 1%	3.75%	1.2%
Actual revenue collected per student	$3,750	$ 96
State government share or supplement	–$2,250	+$384
Average residential tax bill	$1,875	$120
Average residential tax bill before power equalization	$750	$600
Change in residential tax bill due to power equalizing	+$1,125	–$480

one-time change in taxes that is expected to be permanent is easy. The change in property value can be estimated by dividing the change in taxes or benefits by the appropriate discount rate. For example, if we assume that an investment in similar risk property would earn 10%, we simply divide the change in net benefits from property ownership by .10 to determine the change in property value. Let us calculate the changes for average residences in our example districts.

In District A we observe an increase in taxes for the average residence of $1,125. This is a decrease in net benefits from ownership of $1,125 per year. If we assume a discount rate of .10 we can anticipate a fall in the market value of the average house in District A of $11,250 (–$1,125/.10 = –$11,250), from $50,000 to $38,750. At the 10% discount rate, the decrease in value of $11,250 is exactly equal to the amount that would have to be invested at a 10% return to pay the increase in taxes. While property values are falling in response to the tax increase in District A, property values would be rising in response to the fall in taxes in District B. In Table 2 we observe that the average residential tax decrease was $480. If residents in District B also have a 10% discount rate, we can estimate that the value of the average residence would increase by $4,800 ($480/.10 = $4,800) to $14,800.[11]

The new market values calculated from the initial tax changes, however, would not be equilibrium values. This is because at the lower market value the tax bill on property in District A would decrease, and the tax bill in District B would increase in response to the increase in market values there. After several rounds of tax changes and market value adjustments, we would anticipate a new equilibrium for average residential value in District A of $41,700 with a tax payment of $1,564. In District B, the new residential value would approach $14,285 with a tax payment of $171. Power equalizing would have resulted in a reduction in market values and an increase in taxes for residents in District A and an increase in market values and decrease in taxes for residents in District B. These changes in taxes and property values, however, would be one-time effects on old residents.

New residents moving into District A would view the reduction in housing prices ($50,000 to $41,700) as being exactly offset by the higher tax liability they face ($750 to $1,564); for new residents there is no net change in their welfare or real-wealth position from a pre- to a post-power-equalization situation. The cost of the change was borne totally by property owners at the time of the tax change. What about new residents in District B? New residents buying into District B would expect to pay an average of $14,285 for their house, accompanied by a future tax liability of $171 per year. While new residents have lower taxes than old residents did before power equalizing, for the new residents the lower taxes are exactly offset by the increase in the price they had to pay for their house. Thus, new residents in District B are no better off in terms of welfare or real wealth than old residents were before power equalizing. However, the property owners at the time of the equalization have made capital gains.

Can we say that power equalizing has raised the *fiscal capacity* of the low-income district? As new owners move into District B, their ability to spend on education will be limited by the fact that their housing payments are higher. New residents in District A will face higher taxes but be able to spend more because their housing payments are lower. *Fiscal capacity* has not changed because costs of residing in the district have not changed.

It is likely that in the short run the effects of tax changes on existing homeowners will lead the residents in District A to spend less on education. They paid $50,000 for their houses, expecting taxes to be $750; if taxes suddenly rise to $1,564, they may decide to tax and spend less. On the other hand, homeowners in District B who paid $10,000 for their houses, expecting to pay $600 annually in taxes, may find that they want to expand financing for education when their new taxes come in at only $171.

Much more important than wealth effects for influencing changed expenditures on education, however, are *price effects.* Originally in District A, each $1 per pupil spent on education cost each homeowner $0.50, because residential valuation represented only half of the wealth of the district. Under the new situation each $1 of educational expenditure costs each homeowner $1.39; a price increase of $0.89 per dollar spent on education, or 178%. In District B, each $1 per pupil of expenditure originally cost $1.25 per homeowner. Now with equalizing, each $1 of expenditure costs only $0.33, a price decrease of $0.92, or 74%.[12] With these relative price changes we anticipate reduced spending on education in District A and increased spending in District B. This is the effect school finance reform advocates are looking for. The changes in the level of education provided, however, may best be viewed as a *consequence of relative price changes* of education, not relative wealth changes or equalization of fiscal capacity.

The data selected to illustrate power equalizing in the above example represent an extreme case where the rich district contains both rich people and highly valued business property and the poor district contains low-income families with low-valued residential property. Thus in the example, the changes in capital values were progressive, that is, the rich lost and the poor gained. When there is almost zero correlation between high-wealth districts and high-income residents, capital gains and losses from a switch to a power-equalization formula would simply generate windfall gains and losses randomly. The price effects would encourage increased spending among formerly low-wealth districts (which may well contain middle- or high-income residents) and decreased spending among wealthier districts (including central cities). Thus the price effects would result in more equal expenditures for education, while capitalization of tax and expenditure changes into property values would result in large windfall gains or losses to current property owners, and no net welfare change over the pre-power-equalizing situation in the long run. Residents in poor districts will look wealthier because their property will be more highly valued when taxes fall, but this wealth is a mirage for a new resident who pays a higher price for a house and hence has higher housing payments. Residents in rich districts will look poorer because their property will be less valuable, but this is also a mirage, for new residents who purchase their houses for a lower price will have lower housing payments.

SUSPECT CLASS: PROPERTY VALUE OR PRICE

To an economist it appears anomalous that a suspect class is defined as a group of people who pay a lower price for a house and higher taxes in relation to another group which pays a higher price for a house and lower taxes. Property value is not really a meaningful measure of wealth or fiscal capacity, and the use of wealth to define a suspect class is probably best reserved for application to individuals or families. This conclusion was also reached by the U.S. Supreme Court in San Antonio Independent School District v. Rodriguez (1972).

While property value is not a good measure of wealth, however, property value may be directly correlated (negatively) with the price of a public good or service. If the unequalizing effects of price differences are to be eliminated, a program such as district power equalizing will accomplish that objective. The "suspect class," however, is comprised of persons paying higher prices for the good and has nothing at all to do with the richness or poorness of individual families. This approach to suspect class has less precedent than trying to tie district wealth to the series of previous cases identifying poor individuals as a suspect class, but to an economist it is much more meaningful.

SCHOOL FINANCE REFORM AND THE POOR

With very few exceptions the current thrust in school finance reform is advertised as if it were designed specifically to help poor people achieve levels of education.[13] It must be emphasized that programs based on school district wealth or property value *have nothing to do with individuals who are poor* because there is essentially no correlation between school district wealth and family incomes. Poor people who reside in low-wealth districts will benefit from the capital gains associated with tax reductions and will spend more per student for education because of its lower price. Poor people who reside in high-wealth districts will bear the burden of capital losses associated with tax increases and will spend less per student for education because of its higher price. New residents in either kind of district will not bear capital gains or losses, but there will be less difference in expenditure levels among districts. School finance reform is simply not related to poverty or poor people; it is related to unequal expenditures for education. If aid for poor people is desired it will have to be attempted through a different approach.

IMPLICATIONS FOR LOCAL GOVERNMENT FISCAL REFORM

While the school finance reform movement has based its arguments on eliminating wealth as a determinant of the quality of education a child receives and not necessarily reducing the variations in expenditures per pupil (Coons et al., 1970: 438), a closer look at the issues contained in reform proposals indicates that the outcomes would be more equal expenditure levels for education due to price changes and not any change in wealth in terms of fiscal capacity or net well being of individuals (except for the short-term capital gains and losses). Thus, school finance reform is more closely related to those aspects of local governmental fiscal reform that have focused on unequal tax rates and expenditures per se as undesirable and less to those aspects of fiscal reform associated with income redistribution or poverty. Let us look briefly at the implications that can be drawn from the examination of school finance reform, especially land rent theory, for other aspects of local government fiscal reform.

INCOME REDISTRIBUTION

Capitalization of the value of government programs or costs into property values greatly reduces the capacity of the states and the national government to redistribute income through programs targeted to small geographic areas. For example, if governmental benefits are provided in a small model-cities area but are not available outside that area, it is likely that the benefits of the programs will be captured in higher property values for landowners in the model-cities area. The citizens who receive the government services will find the value of those services offset by the higher rents they pay to landlords. Citizens who owned their own home and were also eligible for government services would of course benefit directly, but the number of homeowning persons who need income redistribution is relatively small.

Income redistribution to poor people, either through tax reduction or increases in public programs must be tied directly to poor people and not to geographically defined units such as small local governments. Providing aid directly to poor people without regard to their geographic location also may benefit poor people by permitting them to migrate to locations where jobs are available instead of remaining tied to older areas of cities where, because the poor are concentrated, more social services are available (Bish, 1971: chapter 7).

PROPERTY VALUE AND FISCAL CAPACITY

The lack of a correlation between property values and family income–or, for businesses, profits–in small jurisdictions makes the use of property value as a measure of fiscal capacity meaningless if by fiscal capacity we want to get some idea of the ability of individuals or businesses to pay for the support of public programs. Family income and, for businesses, profits are better measures of ability to pay–and even these indicators are not perfect.

GOVERNMENT EFFICIENCY AND TAX COMPETITION

Increased local government efficiency (lower taxes or higher benefits) will be reflected in property values. This may well be one reason why large landowners and real estate speculators are so active in local government politics. From this perspective, tax competition is only one-half of the issue–benefits must also be considered. One should anticipate, however, that a large share of the benefits from locating a new business in an efficient governmental unit will be captured by property owners rather than by the new business. It is also anticipated that competition to attract new businesses will provide an incentive for government officials to provide only programs where benefits exceed costs by the greatest amount. Such an incentive is not undesirable, even though many local government fiscal reform advocates view it as such (Bish and Nourse, 1975: chapter 6).

EXPENDITURES ON SPECIFIC SERVICES

If a unit of output can be designated (such as expenditure per student), increased or decreased expenditures by local governments can be encouraged by *changing the price* of that good or service to individual taxpayers in local units. We would expect any net benefits or costs of such changes to be capitalized into property values and the long-run net welfare of new residents to be unchanged from that of residents in the district prior to the change. Thus, gains and losses of such changes are largely borne by property owners at the time of the change.

CONCLUSIONS

There is considerable confusion in analyses of local government fiscal reform and school finance reform. Much of the confusion occurs because it is assumed that poor people reside in local government units characterized by low wealth as measured by property values. A second source of confusion is the failure to recognize the capitalization of changes in costs and benefits from government programs into property values and the consequent weakness of property value as a measure of fiscal capacity.

When these issues are recognized, two conclusions seem to stand out. First, expenditures by local governments for any particular good or service can be influenced by manipulating the price of the good or service. This manipulation could include lowering prices to increase expenditures, raising prices to reduce expenditures, or doing both simultaneously—as with power equalizing—to achieve more equal expenditure levels. Second, attempts either to alter "fiscal capacity" of local governments or to provide income redistribution via grants to local governments are not likely to be successful because the costs and benefits are likely to be capitalized into property values. Thus, income distribution problems are better treated by focusing on individuals instead of on local governmental units.

These conclusions should be reconcilable with legal approaches. Prices for goods of "fundamental interest" could be equalized so that a spurious element, price difference, is not a determinant of expenditure or consumption levels. Wealth as a suspect class could be confined to the wealth of individuals or families, which is much more meaningful than the "wealth" of a local government when its own actions partially determine that wealth. Finally, income redistribution per se could remain focused on poor people and be determined primarily through legislation rather than through the courts. This approach would clarify some of the issues that are not clear in school finance reform (although I believe it is consistent with the basic arguments in Coons et al. [1970]) and provides an improved framework for analyzing the general issues surrounding local government fiscal reform. It would also make it much easier for scholars from different disciplines to work together to identify programs which can be predicted to achieve the desired results.

NOTES

1. These problems have been stated in their traditional form. This restatement does not indicate that I agree with them, but their validity is not the focus of this paper. For a critique and analysis of these issues—especially the suburban-urban exploitation hypothesis—see Bish and Nourse, 1975: chapter 6; and Bish and Ostrom, 1973.

2. Section I of the Fourteenth Amendment states:

> All persons born or naturalized in the United States and subject to the jurisdiction thereof, are citizens of the United States and of the State wherein they reside. No State shall make or enforce any law which shall abridge the privileges or immunities of citizens of the United States; nor shall any State deprive any person of life, liberty, or property, without due process of law; nor deny to any person within its jurisdiction the equal protection of the laws.

While originally ratified in 1868 to protect the interests of Blacks, like many constitutional provisions the Fourteenth Amendment has become a basis for decisions far removed from—but related to—its original purpose.

3. Analyses of the evolution and application of these concepts are presented in Graham and Kravitt (1972) and Hogan (1974).

4. District power equalizing can also be viewed as a logical extension of "percentage equalizing," a kind of program used in several states. With percentage equalizing the state government guarantees that all school districts can spend a specified minimum per student at a specified tax rate, even if at that rate less than the specified minimum would be raised from within the district. The state makes up any revenue insufficiencies from general funds. For example, a state government could determine that an appropriate tax rate was 2% and that all districts should be able to spend $600 per student if each levied a 2% tax rate. Then any district with an assessed value level below $30,000 per student (the amount necessary to raise $600 at a 2% rate) would receive state supplements that would bring its total revenue per student to $600. Percentage equalizing permits all low-wealth districts to achieve some minimum level of expenditure without excessive tax effort. Power equalization can be viewed as an extension of percentage equalizing where, in addition to supplementing the revenues of low-wealth districts, the state takes revenues above specified amounts away from wealthy districts. The objectives of the two approaches are considerably different, however. Percentage equalizing is designed to guarantee *minimum* levels of expenditure in all districts; power equalizing is designed to *equalize* per-student revenue per unit of tax rate in all school districts (Coons et al., 1970: 163-173; Bish and Nourse, 1975: 308-309).

5. Coons et al. (1970) also apply the concept of district power equalizing to income taxes and to individual families under a voucher kind of program. Neither of these latter applications has received the emphasis of district power equalizing based on property taxes, even though I believe a better case can be made for the income-tax-based alternatives, especially family power equalizing.

6. The existence of different tax rates in different school districts was in fact the basis for a successful challenge to the constitutionality of the school finance system in New Jersey (Robinson v. Cahill, 1973).

7. A correlation of 1.0 would mean that income and wealth were perfectly related; a correlation of 0 would mean that they were totally unrelated. The observed correlation of 0.02 means that, for all practical purposes, income and wealth are *not* related at the school district level in California.

8. A discussion of capitalization can be found in any public finance or urban economics text. An extensive analysis is also found in the law text by Oldman and Schoettle (1974). An illustration and calculation of capitalization is presented later in this article when an analysis is made of the impact of power equalizing.

9. Sections of this analysis including the example are from Bish and Nourse (1975): 302-307.

10. The state could set any fixed relationship between tax rate and expenditures, not just one related to average wealth per student statewide.

11. These calculations have been simplified for the purpose of clarifying the capitalization process. In analyzing an actual change, one would have to take into account two additional factors. First, calculations must be based on the net cost to a taxpayer of a change in taxes and housing prices after taking income tax deductability of taxes and interest into account. The effects of an increase in property taxes, when these factors are taken into account, would be slightly less than indicated in the numerical example, because all property taxes are deductable but only the mortgage interest payments on the reduced housing price are deductable. On the other hand, the numerical example slightly understates the amount of capitalization when property taxes are decreased and housing prices increase. Second, if a home buyer places a value on being able to purchase additional education at a low price (as in District B) this may also lead him or her to be willing to pay a higher price for the house. Conversely, a new higher price for education may lead a buyer to offer less for a house in District A. These effects would reinforce the strength of initial capitalization of tax changes on housing prices.

These examples have also been presented under the assumption of full capitalization of property tax changes into real estate values. Empirical evidence supports an expectation of full or a very high level of capitalization on both owner-occupied and rental housing.

12. One should note that prices in terms of the dollar cost of obtaining education to families in Districts A and B are not equalized. Only the dollars raised per *tax rate* have been equalized under power equalizing.

13. Coons et al. (1970) make it clear that while they believe low-wealth districts are usually populated by low-income residents, their argument does not depend on any such relationship. They argue that one can view *all* children as poor and that families having to pay higher tax rates (i.e., those in low-wealth districts) for the same level of expenditure on education are disadvantaged whether or not the family itself is poor.

CASES

Boddie v. Connecticut (1971) 401 U.S. 371.
Brown v. Board of Education (1954) 347 U.S. 483.
Griffin v. Illinois (1956) 351 U.S. 12.
Griswold v. Connecticut (1965) 381 U.S. 479.
Korematsu v. U.S. (1944) 321 U.S. 214.
McInnis v. Shapiro (1968) 393 F. Supp. 327 (N.D. Ill.).
Reynolds v. Sims (1964) 377 U.S. 533.
Robinson v. Cahill (1973) 62 N.J. 473, 303 A 2nd 273.
Rodriguez et al. v. San Antonio Independent School District et al. (1972) 377 F. Supp. 280 (W.D. Tex.).
San Antonio Independent School District v. Rodriguez (1972) 411 U.S. 1.
Serrano v. Priest (1971) 95 Cal. Rptr. 601; "Memorandum Opinion re Intended Decision," Super. Ct. of the State of California for the County of Los Angeles, filed April 10, 1974.
Shapiro v. Thompson (1969) 394 U.S. 618.
Skinner v. Oklahoma ex rel. Williamson (1942) 316 U.S. 535.
Strauder v. West Virginia (1880) 100 U.S. 303.

REFERENCES

Advisory Commission on Intergovernmental Relations [ACIR] (1973) Financing Schools and Property Tax Relief–A State Responsibility. Washington, D.C.: Government Printing Office.

--- (1971) Measuring the Fiscal Capacity and Effort of State and Local Areas. Washington, D.C.: Government Printing Office.

--- (1967) Fiscal Balance in the American Federal System. Vol. 2, Metropolitan Fiscal Disparities. Washington, D.C.: Government Printing Office.

--- (1965) Metropolitan Social and Economic Disparities: Implications for Intergovernmental Relations in Central Cities and Suburbs. Washington, D.C.: Government Printing Office.

AVERCH, H. A., S. J. CARROLL, T. S. DONALDSON, H. J. KIESELING, and J. PINCUS (1972) How Effective is Schooling? A Critical Review and Synthesis of Research Findings. Prepared for the President's Commission on School Finance. RAND No. R-956-TCSF/RC. Santa Monica, Calif.: RAND Corporation.

ANDERSEN, W. R. (1975) "School finance in Washington–the Northshore litigation and beyond." Washington Law Review 50: 853-900.

BISH, R. L. (1971) The Public Economy of Metropolitan Areas. Chicago: Rand McNally/Markham.

--- and H. O. NOURSE (1975) Urban Economics and Policy Analysis. New York: McGraw-Hill.

BISH, R. L. and V. OSTROM (1973) Understanding Urban Government: Metropolitan Reform Reconsidered. Washington, D.C.: American Enterprise for Public Policy Research.

BLACK, D. E. (1974) "The incidence of differential property taxes on urban housing: some further evidence." National Tax Journal 27 (June): 367-369.

CHURCH, A. M. (1974) "Capitalization of the effective tax rate on single family residences." National Tax Journal 27 (March): 113-122.

COONS, J. E., W. H. CLUNE III, and S. D. SUGARMAN (1970) Private Wealth and Public Education. Cambridge, Mass.: Harvard University Press.

COWING, T. G. (1974) "Real property taxes, local public services and residential property values." Southern Economic Journal 41 (October): 325-329.

COX, M. (1975) "Equality in educational finance: an analysis of the distribution of taxes and educational benefits by district." Ph.D. dissertation. University of Southern California Center for Urban Affairs.

ECKER-RACZ, L. L. (1970) The Politics and Economics of State-Local Finance. Englewood Cliffs, N.J.: Prentice-Hall.

GRAHAM, R. L. and J. H. KRAVITT (1972) "The evolution of equal protection–education, municipal services and wealth." Harvard Civil Rights–Civil Liberties Law Review 7: 105-213.

HARRISON, R. (1974) "What now after 'San Antonio Independent School District v. Rodriguez'? Electoral inequality and the public school finance systems in California and Texas." Rutgers Camden Law Journal 5: 191-217.

HEINBERG, J. D. and W. E. OATES (1970) "The incidence of differential property taxes on urban housing: a comment and some further evidence." National Tax Journal 23 (March): 92-98.

HOGAN, J. C. (1974) The Schools, the Courts, and the Public Interest. Lexington, Mass.: D. C. Heath.

HUGHES, J. F. and A. O. HUGHES (1972) Equal Education. Bloomington: Indiana University Press.

HYMAN, D. N. and E. C. PASOUR, Jr. (1974) "Real property taxes, local public services, and residential property values." Southern Economic Journal 41 (October): 329-331.

--- (1973a) "Property tax differentials and residential rents in North Carolina." National Tax Journal 26 (June): 303-307.

--- (1973b) "Real property taxes, local public services and residential property values." Southern Economic Journal 39 (April): 601-611.

MAXWELL, J. A. (1969) Financing State and Local Governments. Washington, D.C.: Brookings.

Note (1969) "The right to adequate municipal services: thoughts and proposals." New York University Law Review 44: 753-774.

Note (1972) "A statistical analysis of the school finance decisions: on winning battles and losing wars." Yale Law Journal 81: 1303-1341.

OATES, W. E. (1969) "The effects of property taxes and local public spending on property values: an empirical study of tax capitalization and the Tiebout hypothesis." Journal of Political Economy 77 (November-December): 957-971.

OLDMAN, O. and F. P. SCHOETTLE (1974) State and Local Taxes and Finance. Mineola, N.Y.: Foundation Press.

ORR, L. L. (1970) "The incidence of differential property taxes: a response." National Tax Journal 23 (March): 99-101.

--- (1968) "Incidence of Differential property taxes on urban housing." National Tax Journal 21 (September): 253-262.

PASOUR, E. C., Jr. (1973) "Real property taxes and farm real estate values: incidence and implications." American Journal of Agricultural Economics 55 (November): 549-558.

REISCHAUER, R. D. and R. W. HARTMAN (1973) Reforming School Finance. Washington, D.C.: Brookings.

RIEW, J. (1970) "Metropolitan disparities and fiscal federalism," pp. 137-162 in J. P. Crecine (ed.) Financing the Metropolis. Vol. 4, Sage Urban Affairs Annual Reviews. Beverly Hills, Calif.: Sage.

SABELLA, E. M. (1974) "The effects of property taxes and local public expenditures on the sales prices of residential dwellings." Appraisal Journal 42 (January): 114-125.

SCHWARTZ, M. A. (1973-1974) "Municipal services litigation after Rodriguez." Brooklyn Law Review 40: 93-114.

SMITH, R. S. (1970) "Property tax capitalization in San Francisco." National Tax Journal 23 (June): 177-193.

TREACY, J. J. and L. W. FRUCH II (1974) "Power equalizing and the reform of school finance." National Tax Journal 27 (June): 285-299.

WHITE, R. D. (1974) "School finance reform: courts and legislatures." Social Science Quarterly 55 (September): 331-346.

WICKS, J. H., R. A. LITTLE, and R. A. BECK (1968) "A note on capitalization of property tax changes." National Tax Journal 21 (September): 263-265.

4

New Forms of Involving Citizens in Educational Systems

LUVERN L. CUNNINGHAM

□ THIS IS THE STORY of a sociopolitical invention, born of crisis and sustained by the belief that its missions are larger than crisis. The invention is a citizens' problem-solving third party—the Detroit Education Task Force. Its creation resulted from a shared concern about the effectiveness of the Detroit Public Schools, even severe doubts about continuing the Detroit Public Schools as an institution. The maintenance and extension of the Detroit Education Task Force have depended upon its effectiveness as measured by the institution to whom it owes its allegiance; a configuration of national, regional, and local interests; and the belief of several foundations that it is a noble experiment worthy of their investment.[1]

BACKGROUND AND CONTEXT

The Task Force is a citizens' group served by a small professional staff. It has been charged by the Detroit Central Board of Education with problem-solving responsibilities. Its commitment to problem-solving—as distinct from problem analysis, review, and recommendation—is unique in the history of citizen involvement in education. The Task Force is a third party in many respects. As its three-year history has unfolded, dimensions of third-party activity have become clearer. Those dimensions are the focus of this presentation.

The Task Force is a creature of the Detroit Central Board of Education but not its captive. It is simultaneously intimate with and remote from the Central Board. It is assisted by the good will of the Central Board but not dependent upon it. It works hand-in-hand with its parent agency and supports the Board's mission but is free to differ with it on issues and problems. The Task Force is expected to construct a problem-solving agenda but enlists the participation of school officials in that process. The relationships between the Task Force and the Detroit Public Schools have been marked by affection, misunderstanding, respect, hostility, love, hate, friendliness, and tension—always tension.

As a problem-solving agent, the Task Force occupies unique terrain, e.g., it is legitimized by a public body; it serves an ambiguous "public interest"; it is extremely heterogeneous in its membership; it has a professional staff that is marked by diversity; it is expected to exhibit neutrality, impartiality, and objectivity; and it is viewed as a temporary system at best.

"Public" views of the Task Force obviously are varied. Most citizens of Detroit are unaware of its existence. Teachers and other school employees have developed an increasing recognition of its presence after three years but are generally not well informed about it. School administrators know about it and either tolerate, respect, fear, or disregard it. Central Board members responsible for its creation and extension hold widely varying views about it. Regional Board members (except chairpersons) are largely indifferent toward it. The superintendent endorses and applauds its work but is sometimes inconvenienced by it. Nearby universities are aware of the Task Force and are involved in its work. Media leaders know about it, publicly recognize it, and have been friendly toward its leaders and professional staff. The *Detroit Free Press* editorially endorsed its work in July of 1975 and advocated the application of the problem-solving concept and strategy of the Task Force to other areas of the local government. It is cheered by New Detroit, Inc. (Detroit's local urban coalition) and from time to time receives recognition from other groups, individuals, and associations.

DETROIT AND ITS SCHOOL SYSTEM

Detroit is the nerve center for much of the world's automobile industry. It is also a city marked with decay. The riots took their toll in 1967 and urban blight is everywhere. It is a city with extraordinary crime problems, a city at the center of the heroin traffic in the United States, and a city with the highest homicide rate in the nation. In 1974, 751 homicides were committed within its city limits. That rate is growing—each year the homicide rate is beyond what it was the previous year. The unemployment rate for males between the ages of 16 and 25 is approximately 40% and has been for several years. The economy is tied fundamentally to the automobile industry, and the harsh impact of inflation on this industry foreshadows severe employment problems.

Detroit is the nation's fifth largest city. It is essentially a single industry town, deeply devoted to unions and collective bargaining as the mechanism for distributing income. It has grown rapidly during this century to a peak population of 1,849,568 in 1950. The ratio of whites to blacks has shifted dramatically in the past two and one-half decades. In 1950, the city was 83.6% white, 16.4% black; in 1970, the ratio was 55.5% white and 44.5% black. The school population in 1971 was 64.9% black, 33.3% white; in September 1974, it was 71.6% black, 26.4% white (Thompson, 1974). The Spanish speaking comprise the largest other minority group, but it is not yet a significant political or social force in the city.

The school system is plagued by problems of retrenchment, declining enrollments, inflation, archaic structures and policies, decentralization and community control, public discontent and disaffection, racism, poor pupil achievement, security, inadequate finance, a fatigued work force, bad communications (internal and external), court-ordered desegregation, tired leadership, contending authority systems, and malfunctioning collective negotiations processes, to name a few.

The kinds of problems junior and senior high school principals face in this city are staggering. Those of us in academic circles have little sense of the gut-grinding dilemmas that these people encounter every day. We should be shamed by our impotence as we stand witness to what occurs there.

A few months ago, Bernard Watson, a well known authority on urban education problems from Temple University, made a presentation to the administrators of Region Six in Detroit. He described principals as people "on the point." It was an appropriate military analogy; principals in this city are, indeed, on the point. They face problems of pupil truancy, teacher absence, and the inability to manage the day-to-day affairs of the school. There is drug traffic in the corridors, pimps on the street, and pushers waiting outside. It is an environment marked by despondency and despair.

One of the Task Force staff members recently shared his reactions to a secondary school where he had been involved in the analysis of a special program. He talked about the inability of the school system to keep proper records, to provide transcripts for persons who are going to college. He noted the indifference that school people developed toward young persons. He described how young girls who received failing grades on their report cards were easy prey to pimps. He reported on the aroma of marijuana in school corridors. It was a stark and humbling report. Obviously that description does not characterize all of the junior and senior high schools in Detroit or other large cities, but it is true enough of many of them for this society to be extraordinarily concerned about ways to formulate solutions to those problems.

THIRD PARTIES AND THEIR DEFINITION

Affairs of government have fallen on hard times. The weight of everyday responsibility is frequently too burdensome for many structures of government to carry effectively. Consequently, administrators, officers, and legislative bodies are turning to the "people" for help. The organization of systematizing of added public input into governmental processes takes many forms and proceeds under several names. "Task Force" is currently very popular; "commission," "committee," and "panel" are also widely used. In each case, citizens are asked to assist with the affairs of government or the affairs of administration.

The practice is not new. It has been going on for some time and at all levels of government. In the late 1960s, the Kerner Commission was established by President Johnson. Its charge was to look extensively and intensively at public disorder. The commission rendered a report which received wide publicity, and theoretically that report was to affect decision-making at several governmental levels and among many administrative bodies within government. It was also expected to have some impact upon the attitudes of

American citizens. The Kerner Commission report did receive wide distribution; it was the subject of discussions on campuses and within public forums across the country. Although there is no adequate way to measure its effectiveness, the commission did exist, did prepare and disseminate a public report, and did attract some measure of public note.

The Kerner Commission was unusually "public" in its behavior. Its membership and leadership were well known. Another task force also appointed by President Johnson stands in sharp contrast (Kearney, 1967). It was a special group named in the mid-1960s to examine the needs of public education. This "third party," a 12-member group, also worked intensively upon its charge. It, too, produced a document but rendered the report directly to the president and his executive staff. That report led to the legislation known as the Elementary and Secondary Education Act of 1965. Few people knew then, or are aware now, of that presidential task force. Its report was never made a public record. Few people know its membership even now though it included very prominent Americans. Its product was translated immediately into educational legislation —in fact, the most far-reaching in modern times. Fewer than 90 days passed between the submission of the committee's work and the passage of ESEA.

The term "third party" has a number of meanings. Michael (1973) uses "third sector" to describe structures and processes akin to those reported here. Wellington and Winter (1971) apply the term "multiparty" to an emerging collective bargaining need in decentralized school districts. In this case, the multiparty would include representation from decentralized subunits (regions, districts) as well as the central school authority in labor relations with a centralized labor unit.

The Education Task Force is a "third factor" or "third force" in decision-making, respecting the interests of students and parents and monitoring the work of the Central Board of Education and the professional and nonprofessional work forces of the Detroit Public Schools. Its independence is protected through nonpublic sources of support, an annual review of its role and function, and reasonably low-profile work habits relative to the community. The parameters of its problem-solving effort are not limited to Board of Education zones of responsibility. It can work through other governmental subdivisions and levels where those decision points are essential to problem-solving.

There are other third-party problem-solving groups working on education problems in large cities. The San Francisco Public Schools Commission is one example. Similar groups are at work in Los Angeles and St. Louis. In each case the commitment is to problem-solving, to hardheaded examination of the major difficulties plaguing the schools, and to doing something about them. Most problems have been studied, nearly to death, but they persist. Thus, the dogged determination of citizens giving unstintingly of themselves to solve problems is unusual in local governmental affairs.

THE DETROIT TASK FORCE AND ITS FUNCTIONS

Detroit is a much-studied city. Its libraries are filled with documents produced by study groups, many of which involved citizen bodies and focused upon education. The Romney Report of 1958, several hundred pages in length with dozens of education recommendations, and a citizens' report of 1973, with more of the same, rest on the shelf. There was another report in 1968 and still another in 1971. Each of them involved citizens and, in many cases, professional consultants. Almost none of their recommendations have been implemented.

Some have been marginally successful, but most did not reach the expectations held for them when they were launched. This has been especially true in large cities. In most cases the practice has been to do an intensive analysis and write an elegant final report filled with recommendations. In passing those reports along to boards of education, the assumption was that the recipient public bodies contained within themselves the strength, insight, wisdom, and capacity to implement massive recommendations for change. That proved to be a false assumption.

The leadership of the Detroit Education Task Force, in concert with the leadership of the school system, decided that the design and mission of previous citizen groups had proved inadequate. Thus the philosophy, the performance, the processes, and the practice of this Task Force were to contrast sharply with previous citizen efforts.

In the autumn of 1972, the Detroit Public Schools were on the threshold of bankruptcy. They were facing a $73 million deficit that threatened to shut the school system down by March of 1973. Superintendent Charles J. Wolfe recommended to the Board of Education in October of 1972 that a new citizens' committee be established.[2]

Because of the political and social climate in Detroit, each educational issue becomes a political issue, each educational problem becomes a political problem. Even choosing the membership and leadership of the Task Force becomes controversial. Board discussions about membership have been vigorous and sometimes last for several weeks. Proposals relative to the number of Task Force members to be named initially ranged from four to 2,000. The number of members was resolved at 57 in 1972, and it was decided that the Task Force would be headed by co-chairmen, a very important decision by the way. Each Task Force member takes part in one of three internal working committees–the Organization and Management Committee, the Finance Committee, and the Education or teaching/learning Committee.[3]

As indicated earlier, the Task Force is unique in several ways. First is its commitment to problem-solving. Second is its serial approach to problem-solving, which is at the very heart of its work. Problems are identified, solutions sought, and recommendations forwarded to the Board of Education–one at a time. But members do more than simply review, analyze, and advance recommendations. The Task Force stays with a problem until, in its judgment, it has been satisfactorily resolved. There are no "final" reports replete with large numbers of recommendations to be forwarded to the Board.

The Task Force works on an intimate, day-to-day, problem-by-problem basis with an established procedure of formal monthly reports to the Board. Each month one or two recommendations are advanced complete with documentation and attention to problems of implementation, and in some cases cost figures. And the Task Force assists with implementation of its recommendations in creative ways.

Detroit is literally seething with citizen activity in education. The Task Force is only one example. The school system was decentralized by an act of the State Legislature in 1970. In January of 1971, eight regional boards of education were established that preside, in many cases, over a sector of the city that is larger than smaller cities in the state of Michigan. The enrollments in these regions range somewhere between 26,000 and 50,000 students.

Each of the regions has a powerful advisory council that relates to the regional superintendents and the regional boards of education. These councils are aggressive in working with faculty members and principals in defining policies at the building level. They participate often in the choice of the principal and, in more and more instances, are taking part in their removal.

In addition to this network of participation, there are many other committees with interest in education that relate to and impact upon the educational system in this city. Recently the Board has been relating to two new and potentially powerful citizens' groups. One was created by the Board itself to assist with the preparation of a plan to desegregate the schools. The other was named by Judge Robert E. DeMascio to monitor his desegregation court order. The appearance of the monitoring group has produced some difficulties from the Board's perspective, since it now must respect the work of still another external force designed to assist the district.

In sum, the Education Task Force works closely with the Central Board, the Superintendent of Schools, and the executive staff, and is an experiment in third-party problem-solving. The Task Force stands in an interesting posture vis-à-vis the constituents and clients of the school system on the one hand and the organized profession of teachers and administrators and the Board of Education on the other. From its external vantage point, the Task Force works in the interest of the clients of the schools through patterns of collaboration. As a consequence, a number of things have been learned about the functions of third parties as problem solvers.

PROBLEM-SOLVING FUNCTIONS OF THIRD PARTIES

A basic function, consistent with the philosophy of the Task Force, is that of *problem identification, problem definition,* and *problem-solving.* Not all educational problems are self-evident, nor are the problem definitions advanced by school systems necessarily accurate. Throughout its history the Task Force has been working, even straining, to refine its problem-defining capacities. Related to this particular need is a way of choosing among a series of problems those that warrant priority rankings. In its planning the Task Force has been trying to place priorities on what it will be about. The selection of specific problems involves Task Force leaders, the Steering Committee, and working committees. The ultimate choices are the product of review processes involving members, professional staff, and, in most cases, representatives of the school system.

Not all problems identified yield easily to acceptance or to inclusion in the work agenda of a committee. For example, it has become increasingly apparent that the school system has two strong authority systems that are contending for school system control. One of these is the established, traditional, bureaucratic authority structure; the other is the Detroit Federation of Teachers. The two structures coexist. Teachers have loyalty to each. It is evident that little constructive change can take place without the collaboration of the two authority systems. Yet the Task Force has been unable to find ways to approach this problem.

A second extremely important third-party function is that of *convener.* Persons significant to a particular problem are frequently invited to participate in problem analysis. In the absence of the Task Force, those persons would usually not engage in conjoint activity. Persons from disparate sections of the community (and school system) are assembled to examine issues which reflect upon the Board of Education and its zone of responsibility. Individuals are convened, too, to reflect on issues that extend beyond the boundaries of School Board authority.[4] Although the Task Force association is primarily with the Board of Education, it is free to make recommendations to the City Council, the State Legislature, and other agencies and institutions where problem-solving resources reside.

Related to the convening activity is a third function: *providing a forum for the intelligent review of prominent educational issues and problems.* As an example, in the early autumn of 1973, the Task Force examined the issue of accountability in some depth. It was not selected as an area for problem-solving, but the Task Force produced a position statement on accountability that was made available to interested community groups, to the public school system in Detroit, and to the State Superintendent of Public Instruction.[5] Persons were convened to examine the paper and share in considering how the Detroit Board of Education could proceed to develop an accountability plan for the Detroit Public Schools.

There are serious problems today in sharpening issues that have severe implications for institutional clients and the professionals who serve those clients. The Task Force has sought means and formats to be applied to improving lay and professional understanding of high stake issues. To that end the Task Force organized a series of colloquia, providing for the focused examination of issues significant to the Detroit Public Schools. The colloquia were designed as short-term, intensive learning experiences for laymen as well as for professionals. The topics selected were of substantial and usually transcendent interest. Each topic was chosen for its potential to begin the public review of issues of significance to laymen and professionals alike.

A fourth function is as a proxy for *disparate community interests.* The heterogeneous membership of the Task Force reflects the sentiments, values, and beliefs of many segments of the broader community. The Task Force deals through proxies with that heterogeneity as it reviews and reacts to proposals for change in the Detroit Public Schools. The Task Force is a mediator of the public interest, a translator of preferences into policy directions, a filter of community feeling, and an auditor of the public interest. As the months have moved along, it has become apparent that there is substantial significance to proxy activity. It is becoming a fundamental part of what the Task Force is all about.

For years, in Detroit and other major American cities, there were members of boards of education who were able to establish linkage with important sectors of public influence. Board members, often white but not limited to whites, could pick up the telephone and contact individuals who ranked within the top echelons of the city's power structure or, indeed, the power structure of an entire state. They may not have been top influential citizens themselves, but they could tie into the influence structure of the community. Board members now, however, are not only not in those power pyramids but they usually do not have access to them.

In many of our cities, school board members are recent arrivals into policy-making posts. They have no legacy of community leadership to build upon, no way of tying into the influence structure. They are isolated, and their personal isolation contributes to institutional and power isolation. School Board membership in Detroit is now available to 45 citizens. Before 1970 there were seven members on a Central Board. Short of the Common Council, serving on school boards is about the only arena for public service. There are no training grounds or few steps on the leadership ladder before people become Board members.

Many members of the Detroit Board(s) are persons of modest means. They are by and large dedicated but inexperienced. They find their responsibilities after election awesome and overwhelming (Nystrand and Cunningham, 1974). Some Board members are at ease in relations to their constituencies, but others are not. There is a remarkable transformation that takes place when individuals cross the threshold from candidate to incumbent, from campaign rhetoric to Board room accountability, from the critic to the criticized. Feelings of aloneness are apparent. Not only do Board members find it hard "to deliver" to the satisfaction of their own friends and supporters, but they are cut off from other sources of support. The Task Force helps fill this void.

A fifth function of the Task Force is especially critical. *It is the linkage function.* The Task Force stands as an *important channel to sources of power and influence.* The Task Force membership is extraordinarily heterogeneous. It is, in fact, a microcosm of society in Detroit. There have been several bank executives, the Mayor of the City of Detroit, leaders from a variety of grass-roots organizations, the Governor's top aide, several members of the House of Representatives and the Senate, militant Blacks and Chicanos, the Chairman of the Senate Appropriations Committee, the Speaker of the House of Representatives, the Chairman of the Board of the Chrysler Corporation, a delegate from the Office of the State Superintendent of Public Instruction, civil rights leaders, former School Board members, parents, a doctor, and a couple of lawyers. There have been persons from the far right and the far left. There have been Republicans and Democrats, women and men, rich and poor. There have been sharp religious, ethnic, and racial differences–sets of diversity that generate strengths. Each year the membership changes but the representativeness remains.

Almost daily, the Task Force or its professional staff participate in linkage activity, more often than not with school officials. Within a matter of minutes, liaison can be established in two directions. Grass-roots sentiment can be tested about an issue or the perspectives of those thought to be in the power structure can be assessed. Within a short time, leaders can be mobilized from the automobile industry–especially the "Big Three" in the City of Detroit–the unions, the media, the mayor's office, or a network of activity can be established, via telephone, involving representatives of state government.

A sixth function of third-party mechanisms is as a *legitimizer of goals and directions for the school system.* The Education Task Force has forcefully supported the objectives of the Detroit Public Schools. It has worked to reinforce and mobilize resources for the achievement of goals and action programs relating to those goals. For example, the implementation of a Report of the Superintendent's Committee on Achievement has been a major focus of Task Force efforts. A school system committee formed by Superintendent Wolfe produced a report on achievement that was made public in March of 1973. The Task Force examined the document, found it to be of good quality, and endorsed it strongly. From the Task Force's perspective, it was a solid piece of work by a public school committee and, in many ways, has served the Task Force as a road map and a compass for some activities in the area of learning outcomes.

Still another function is *leadership.* Leadership in this context often takes the form of seizing the initiative, moving into a vacuum, and setting in motion a set of events or activities in response to a need. Today's society has serious leadership problems. Constructive leadership is absent in far too many settings. There are few leadership models in our midst that warrant emulation, either in the public or private sectors, and few that reflect understanding of the leadership requirements of settings marked by excruciating demands upon their leaders.

Exhausting inertia exists in large bureaucracies, which imposes exceedingly difficult expectations on those who choose to lead. In fact, there are genuine concerns about the "limits of leadership." How far can an individual or a group "move" an institution or a society? Large bureaucratic structures are so constraint ridden that leadership appears to be an anachronism, especially if there is an expectation that leadership will emerge from within.

The third party has a genuine advantage, at least initially, of meeting leadership needs. It can alter the existing configuration of forces, it can ask old questions in fresh ways, it can raise new questions, and it can convene groups for their examination. The Task Force works with and through established governments but has independence, a life of its own. Furthermore it has an impartiality about it that protects its independence. It can affect the lives of people and institutions as a consequence of the neutral "turf" that it occupies. And it is from that zone of impartiality that it earns respect and, consequently, followership.

TASK FORCE WORK HISTORY

Out of necessity, the Committee on Finance began its work immediately. The day after the Task Force was established (January 4, 1973), its members went to work with legislators and public officials at the state level as well as with local school administrators and the Finance Committee of the Central Board. The Task Force decided to try to keep the schools open for a full 180-day school year. The decision was preceded by sharp debate about the wisdom of that attempt. Many members of the Task Force felt that the schools should close down on March 15, 1973, in part because of the budget deficit but more significantly because of dissatisfaction with the quality of the schools. The closure alternative was abandoned, however, and the Finance Committee set about the search for ways to fund the school district's enormous debt.

Meetings were held in Lansing with representatives of the Governor's office, the State Treasurer, the State Superintendent of Public Instruction, the head of the municipal tax office, delegates from the State House of Representatives and the State Senate. After lengthy meetings, two special pieces of legislation to fund the debt were fashioned and moved through the State Legislature. These were Public Acts I and II of the 1973-1974 session. And they called for interpretation on their constitutionality by the Michigan State Supreme Court.

Task Force leaders expected that the court would respond quickly, but it did not. There was a five-month delay. The waiting period produced an enormous anxiety and tension within the Task Force and Detroit Public School officialdom. There was great concern about constitutionality, and delay in the determination was disturbing. If either half of the two-part package was found unconstitutional, the plan for deficit financing would fail. The Finance Committee would have to go back to the drawing board and find an alternative way of financing the debt.

In mid-September, an emergency meeting of key state officials and Detroit representatives was held in Lansing. Twenty intensely busy citizens and public officials sat around a massive conference table in a state office building for a morning, trying to decide how to approach the State Supreme Court for the interpretation on constitutionality. Several persons present were friends of justices on the court, but none wanted to approach them on a personal basis. The Governor was reluctant to put pressure on the court and so were members of the State Legislature, the State Treasurer, the State Superintendent, the General Superintendent of the Detroit Public Schools, Citizens Research Council of Michigan, and the Task Force. It took three hours of discussion to decide how the approach to the court was to be made.[6]

The group finally found a way of fashioning a letter, which would come from the Governor's administrative board, directed to the State Supreme Court asking why the interpretation had not been made. The letter was drafted and sent. Within ten days, a decision was in hand. The Supreme Court found half of the package constitutional and the other half unconstitutional. The Task Force and the school system were right back where they started from: $73 million in debt and six months of work (and waiting) down the drain.

The Finance Committee went back to work immediately. Persons were convened from the banking community (most of whom were Task Force members) on several occasions to find a way to cover the deficit. The only way was through selling bonds. That solution had been abandoned in early 1973 because of the poor bond risk rating of the city and the Detroit Public School District. In order to improve the ratings of these two independent governments and eventually to sell bonds at a reasonable interest rate, a full day's visit to Detroit was arranged for representatives of Moody's and Standard and Poor's. Each of these bonding houses establishes ratings upon governmental jurisdictions in the United States.

On the day of bonding house visits, a luncheon was arranged by the Task Force at a private club in downtown Detroit. New Detroit, Inc., Task Force, school district, and civic leaders were present to speak on behalf of the city and its schools—indeed, the power structure of the business, educational, and financial communities participated in this event. As a result of that luncheon, the Detroit Board of Education received a substantially improved bond rating. It resulted in the purchase, by insurance companies and banks inside and outside of the city, of several million dollars in deficit bonds with the assurance that the district was a reasonable risk. The school system's interest rate was much improved, which will save the taxpayers several million dollars in interest. The intense fiscal crisis was relieved.

To be sure, the financial condition of the system is still desperate, even though it now operates with a balanced budget. But there is improved public confidence in the fiscal management system and in the Board of Education's public responsibility.

The work of the Task Force's Organization and Management Committee began with an intensive review of the administrative structure of the Detroit Public Schools. On August 14, 1973, an extensive report was presented to the Central Board of Education with detailed plans for reorganizing central and regional administrative structures. The report itself was prepared with the close collaboration and participation of school officials. The Superintendent of Schools and his staff began the implementation of these recommendations shortly after they were received, but it took nearly eight months for them to be put in place. One of the recommendations called for a Department of Finance with a new fiscal officer. That person was identified with Task Force help and began immediately to develop a new fiscal information system and better coordinated departments and divisions of payroll, personnel, and budget, plus improved applications of computer technology to fiscal problems within the schools.

A new Division of Educational Services was created and a new configuration of supporting divisions and administrative units was achieved. A new administrative cabinet was formed to establish a more intimate linkage between regional administrators and central administrative staff. Several other offices were realigned and tied in different ways to the Office of the General Superintendent. An Office of Superintendency was created with the General Superintendent and the Executive Deputy serving as a team. As indicated before, most of those recommendations have been implemented.

More recently this committee focused on warehousing, delivery of books and supplies, and use of space in the School Center Building. It endorsed strongly a joint Task Force staff and school system developed proposal for the massive retraining of administrators and supervisors in the system.[7]

The Education Committee of the Task Force has been confronted, in many ways, with the most challenging assignment of all. Its members have stepped into the quagmire of concerns about the teaching/learning process itself. During 1973-1974 the leadership of this committee was in the able hands of a black gynecologist from a prominent Detroit family. She is Dr. Ethelene Crockett, an impressive person in every respect. She is exceedingly intelligent, energetic, charismatic, and aggressively outspoken about her displeasure with learning outcomes in the Detroit Public Schools.[8]

Members of her committee worked through an exhaustive agenda. Reading and communication skills were the prime focus of attention. The committee made recommendations relative to reading, multicultural and multiethnic education, and health services. Studies and recommendations about severe forms of alienation and a newly proposed LinC (Learning in Community) Semester were advanced, too. (The LinC Semester would allow secondary school students to have one full semester of productive work in the community for which they would receive credit.) More recently, the Education Committee examined counseling and guidance services and developed a set of recommendations regarding cable television and community information systems.

The program recommendations of the Detroit Education Task Force developed during 1974-1975 were incorporated in large measure in the desegregation court order issued by Judge Robert E. DeMascio. Specifically the areas included in the court order were vocational education, counseling and guidance, reading, bilingual and multicultural education, and management training for school administrators. The order mandates the implementation of these programs. The implementation of those and other features of the court order are monitored by a special citizens' committee created for that purpose.

The track record of problem-solving within the Task Force committee structures has been satisfactory. Members, especially Task Force leaders, have given generous amounts of time to committee efforts. But the Task Force work has not been limited to committee sponsored activity.

A UNIQUE EXAMPLE OF THIRD-PARTY ACTIVITY

The 1973 teachers' strike in Detroit was devastating. The bitterness, the heartache, the divisiveness, and the near violence that it produced will scar the institution for a long, long time. It divided administrators from teachers, professionals from laymen, and produced extraordinary alienation on the part of persons working at the classroom level.

The bargaining impasse of 1973 grew out of an inability to deal with an "accountability" question. The central administration insisted that it be placed on the bargaining table. The union not only rejected the issue of accountability but all proposals for achieving an accountability system. The stalemate nearly tore the school system apart.

In many ways the students of Detroit paid the heaviest price. Their school year was extended into the hot, humid Detroit summer—until July 12th. Students attended classes in non-air-conditioned buildings with extreme discomfort. They stayed away in droves. Attendance dropped markedly, which in turn jeopardized state aid. At school's end in July, teachers were tired, administrators worn out, students discomforted, and parents angry. The prospect of another strike was imminent. The Board and the union had been unable to even begin the 1974-1975 bargaining process, much less demonstrate progress in their negotiations.

Because of this vacuum, several people, including the author, became involved in an attempt to work out a package solution that could be introduced to both parties. In effect a special problem-solving third party was put in place. It had been recognized within the Task Force months earlier that bargaining in the public sector was archaic and cumbersome. "Bargainers in public education are still in kindergarten," said one leading Detroit labor leader. The public interest is often poorly served as a consequence.

Much of the leadership for this special endeavor came from private-sector labor professionals, but the day-to-day direction came from the President of New Detroit, Inc., Lawrence P. Doss. The third party was composed of eight leaders from the AFL/CIO, the United Auto Workers, the Mayor, and representatives of New Detroit and the Education Task Force. The labor leader members of the group worked around the clock on a package settlement that would be acceptable to both the union and school officials. The package included wages, the accountability issue, and a modification of class size. It also postponed negotiation on an issue of "teacher residency" until October and swept off the table approximately 100 issues that had been placed there by both parties, postponing action on those until the autumn of 1974.[9] The settlement included the requirement that the bargaining process begin for 1975-1976 in October, which would allow the budget to be developed for the next year with a better indication of the fiscal demands likely to be made upon that budget to meet salary demands from the several bargaining units.

It was an extremely sensitive and potentially dangerous third-party intervention. Had it failed it would have been a conspicuous failure, jeopardizing the reputation of the Task Force. A complete scenario of the events that took place during those several days is of textbook proportions and would require more space to recount here than is available. Close to the end of the negotiations, meetings were held with the top editorial staffs of the metropolitan newspapers and public affairs executives of radio and television stations to prepare them for the announcement in the event a settlement package was achieved. These meetings took place on a Wednesday. The State Mediation Board was approached and agreed that two mediators would present the package formally to the bargaining teams of the Board and the union on Friday afternoon. A citywide press conference was held in the Mayor's office the next morning (Saturday) during which representatives of the Task Force, New Detroit, Inc., the Mayor, and other civic leaders endorsed the package.

Associations with media leaders were exceedingly interesting during this period. They were unanimous in their belief that the paralysis of negotiations required third-party intervention and that the work of the third party had to remain confidential until the appropriate time. They were likewise unanimous in their judgment that another teachers' strike would not only destroy the school system but the City of Detroit as well.

It was a sensitive, touch-and-go situation. Premature reporting would have destroyed the entire effort. The school system, the city, and the state would have been plunged right back into the quagmire. And the prospect of constructing a new agreement out of such rubble was dismal indeed. Thus, a trade-off with the media was exacted: the third party gave them all of the information about the package, the timetable, and time for their editorial writers to prepare editorial statements for Saturday, Sunday, and Monday editions and broadcasts in return for keeping the story under wraps until Saturday.[10]

As it turned out, this story had a happy ending. Two representatives from the State Mediation Board presented the proposed package to a combined meeting of union and School Board representatives at 4:00 p.m. on Friday, July 12th. The Board of Education and the superintendent reviewed the proposal as did the Executive Committee of the Detroit Federation of Teachers. There was some static on both sides, and for three or four days there was fear that one side or the other would reject the package.

Reconciliation of differences regarding details and the translation of the package into contract language took place. Union leaders submitted the agreement to DFT members by mail referendum. The ballots were tabulated on August 2nd. The vote was overwhelmingly in support of the package, and the strike was avoided.

The success of this effort produced a ground swell of favorable public sentiment. Newspaper, radio, and television editorials were generous in their applause for teachers and school officials for settling. The ad hoc third party was singled out, too, for special praise. This third party was not the brainchild of the Education Task Force alone, nor was it quarterbacked by the Task Force. It was, however, an intensive effort totally in keeping with the problem-solving philosophy and commitment of the Task Force.

SUPPORTIVE STRUCTURES

The formal relating of the school system to the Education Task Force is through a school system appointed liaison officer. Ms. Aileen Selick, an exceptionally competent administrator, has served in this capacity through the entire history of the Task Force. It is a delicate, sensitive role requiring respect and confidence from both partners in this collaborative arrangement. Considerable information is transmitted through the liaison office. The General Superintendent and the leadership of the Task Force are dependent upon the competent performance of this role.

The Task Force staff has included professionals from the Citizens Research Council of Michigan headed by a gifted director, Dr. Robert Pickup. He and his colleagues have taken the lead in the analyses of problems of finance and fiscal administration as well as in administrative reorganization. Their staff work for the Finance Committee and the Organization and Management Committee enabled these subgroups to progress rapidly. Their competence highlights the significance of solid staff work to support the efforts of thoughtful laymen involved in third-party activity.

The institutions of higher education in southeast Michigan have been involved in important ways, too. George Gullen, the President of Wayne State University, is chairman of an important Advisory Committee from Higher Education. Several postsecondary institutions have joined in the common objective of assisting the Detroit Public Schools. Individual faculty members, graduate students, and administrators from these institutions have been involved in a broad range of Task Force activities. Their spirit of participation in mutual problem-solving is very impressive. In prospect for the future is a thorough examination of teacher, counselor, and administrator preparation, and the development (in cooperation with the State Superintendent) of a teaching improvement center in Detroit.

During the first two years, some noteworthy developments occurred relative to the joint research interests of higher education and the Detroit Public Schools. A research policy panel was established and a jointly sponsored research seminar was held. New formats for collaboration have been discovered that promise to improve the utilization as well as the generation of knowledge about education.

Through the assistance of Ohio State University, a "Superintendent's Issues Seminar" has been established. It was designed on the basis of a year's work in Columbus, utilizing Harold Lasswell's (1971) concept of the "decision seminar." After several months of Detroit experience, it appears that the system's leadership is about to gain an important new teaching, learning, and planning tool.

The Detroit Decision Seminar at Ohio State during 1973-1974 was of substantial importance to the Education Task Force work in Detroit. Harold Lasswell and Richard Snyder were key figures in the development of the seminar, as was Larry Slonaker, a specialist in communications theory and technology. The seminar was conducted in a specially developed, data saturated, "map and strategy" room.

A room was converted at Ohio State into a unique decision context. Tracks were suspended from the ceiling, which allowed the display of data about Michigan, the SMSA, the City of Detroit, and the Detroit Public School System. Seminar participants were immersed in information about the total environment of Detroit. They were aware simultaneously of economic data significant to problem-solving in Detroit, demographic data significant to problem-solving in Detroit, and educational data significant to problem-solving in Detroit. They were aware, too, of the goals that have been adopted for city purposes, educational purposes, southeastern Michigan purposes, and state of Michigan purposes. Participants learned a great deal about data, about the affective quality of data, the misleading nature of certain kinds of data, and the difficulty of synthesizing and integrating complex data from many sectors, each of which has significance in its own terms for a particular policy or decision before the Board of Education.

The Ohio State Decision Seminar was a testing ground for the Superintendent's Issues Seminar in Detroit. The Superintendent's Issues Seminar focuses on the transcendent overarching, meta, large-scale issues confronting the schools. The first issue before the seminar was retrenchment: the no-growth society, zero population growth, declining enrollments in the public schools, plateauing tax bases within the city of Detroit and the state of Michigan, declining levels of employment, the unemployability of young persons within the school district, and the meaning of a fatiguing, declining, constraining, contracting environment for the schools. The implications of these conditions for personnel, programs, use of school facilities, and finance are under consideration.[11]

LEARNINGS

The premises undergirding the problem-solving commitment of the Task Force are essentially sound. Problem-solving in urban education *is possible,* the serial approach *makes sense,* and citizens *can perform* third-party roles effectively.

We have learned that conjoint activity involving board of education members and school officials linked collaboratively with outsiders must be based on trust. We have learned how difficult trust and confidence are to construct and maintain. We have learned that an overdependence on the part of the board of education is dangerous. We have witnessed firsthand the meaning of the "love-hate" relationship and, on occasion, the emotions involved in "détente."

We have learned about effecting change in large bureaucratic structures. We have developed new respect for their glacial nature, for the importance of achieving changes produced by coexisting, contending, and often contentious authority systems. As indicated earlier, in Detroit, the teachers have loyalties to two powerful authority structures. At times, those loyalties flow in response to the traditional, legal, bureaucratic system. At other times loyalties flow to the Detroit Federation of Teachers. No significant change can take place without the laying on of hands by both of these powerful authority systems.

To achieve change, we have learned that multiple and simultaneous interventions must take place. Those interventions must be informed by understandings of the behavior of multiple authority structures.

We have learned about citizen participation. We know that it is possible to construct a genuine learning community through the participation of dedicated citizens. We know, too, that it is possible to achieve remarkable power from difference. It is possible to locate the chemistry that releases human talent for constructive public purposes. We have learned that communication among widely diverse persons is many, many times more difficult than we had assumed. Citizens have widely divergent styles, preferences, and needs for participation. We have also observed that school officials have substantially more comfort and skill in relating to middle- or upper-class citizens than they do in relating to lower-class citizens.

We have learned that much if not most of the energy available in a large bureaucratic enterprise and its governance system is expended on adult concerns, not the concerns of children and youth. The issues in Detroit of residency, labor negotiations, even desegregation, are predominantly adult centered–not child centered.

We have learned that there are vast reservoirs of talent within a large bureaucracy that are unreleased, blunted, or even destroyed because of misperceptions about the application of rules and regulations and because of the narrowness imposed on human performance by negotiated contracts.

We have learned about the incredible gap between higher education and the lower schools. We have discovered new ways to build institutional bridges, to refine patterns of constructive linkage, to overcome the impotence that those of us in higher education possess. The reconstruction of trust is the point of beginning, and what proceeds from that humble beginning is slow and tortuous.

We have learned that large-scale, comprehensive reform of urban educational systems must be considered in large time frames. We believe that the schools should consider a design that would consume a decade. Each day we recognize more fully that there are no short-run options, no panaceas.

We have learned, too, that the key to leadership, both lay and professional, is to strike common cause and to march to the same drummer. We have learned that the issues which are transcendent, meta, overarching are more important than the trivial. We have learned that the transcendent can capture the devotion and the energies of very diverse persons. We have learned that there are energies available within the society in large quantity to be mobilized for noble public purpose. We have learned that to lead is to expend–it is to expend large amounts of intellectual effort (blended with a generous seasoning of affect) in understanding complexity, in dismantling large puzzling problems, in examining manageable parts, and, over time, in making visible progress with the large questions. We have learned that the concept of third parties has been insufficiently explored. But based on our experience, we believe the notion of third party has extraordinary potential, especially in leadership and change terms.

NOTES

1. Each week, for more than two years, the author spent several days in Detroit serving as Executive Director (during 1974-1975 as Co-Director) of a large citizens' committee. It

afforded the opportunity to observe the problems of everyday management and administration within the largest single institution in the state of Michigan. It permitted the author to witness and take part in an intensive citizen effort to solve the problems of the Detroit Public Schools. And it allowed the author to retain his academic ties within a large public university, the Ohio State University. On Saturdays he taught classes in urban administration at Ohio State and on alternate Mondays during 1973-1974, met with 25 major faculty members from across the entire Ohio State University who joined him in a "Detroit Decision Seminar" considering the problems of Detroit. He had the very best of two worlds. In many ways, his experience paralleled that described by Harold Lasswell as the policy science intermediary, the go-between, who operates within the scientific and academic community and participates in the decision process itself.

2. The Board of Education unanimously approved the resolution to create the Task Force. There is an important set of preceding events leading up to the Board's decision. Powerful members of New Detroit, Inc., including the 1972 chairman of the Board of Directors of New Detroit, Lynn Townsend, urged the chairman of the School Board, James Hathaway, to consider still another citizen effort to help the schools. There was sharp resistance to the idea among School Board members in the summer of 1972, but as the fiscal crisis mounted day by day, that resistance ebbed away.

3. Task Force membership was reexamined in the autumn of 1974. Several persons were added and some shifts were made to accommodate changes in the make-up of the Senate and House of Representatives as a consequence of the November elections. Two new representatives from the Latino community were named. In 1975, substantial changes in membership and organizational structure occurred including the appointment of new individuals to the major Task Force leadership positions.

4. For example, the Task Force and the Board of Education shared responsibility for convening a citywide conference on children and youth. The conference idea was initiated by the Task Force.

5. The district was in the throes of a severe teachers' strike at this time. Accountability was such an emotional issue that the Task Force leadership chose not to intervene directly into the strike (and the accountability maelstrom) but decided to examine the question for its own benefit and without specific recommendations to the Central Board of Education. The Task Force in its early history found two issues so emotionally charged that it could not treat them within its own membership, let alone advance ideas in their regard. They were problems of responding to the integration-segregation questions and the examination of the public interest questions of collective bargaining, public sector–education.

6. This was a lesson to the author personally, an individual saturated with naivete about such affairs. His inclination would have been to call the Chief Justice and find out where the decision was. Obviously it was not to turn out to be that simple.

7. An audit of the years of remaining service to present administration and supervisors in the Detroit Public Schools was most revealing. In 1974, there were 35,000 years of service left in that work force–350 centuries.

8. Dr. Crockett became a co-chairman in September, 1974. She joined Alfred M. Pelham and Stanley J. Winkelman. Pelham and Winkelman were equally impressive and dedicated. Their effectiveness as leaders accounted for a large measure of the third party's success through 1975. Similarly, other committee leaders have been unusually effective in maintaining Task Force momentum. Donald Young, a prominent business executive, became a co-chairman in the autumn of 1975 sharing that responsibility with Dr. Crockett.

9. The Detroit Board of Education passed a resolution requiring that teachers, counselors, and administrators live inside Detroit. The resolution was immediately challenged in the courts by the Detroit Federation of Teachers. The Michigan Supreme Court determined that the matter be returned to the system as a negotiable item.

10. As it turned out, the *Detroit Free Press* ran a late Friday edition story on the settlement. It was given to William Grant, the principal *Free Press* education writer, by Mary

Ellen Riordan, the head of the Detroit Federation of Teachers. It set the stage nicely for the press conference in Mayor Young's office on Saturday morning.

11. The data displayed in the map and strategy room at Ohio State University were reproduced for the Superintendent's Issues Seminar in Detroit.

REFERENCES

KEARNEY, C. P. (1967) "The 1964 presidential task force on education and the elementary and secondary education act." Ph.D. dissertation. University of Chicago.

LASSWELL, H. D. (1971) A Preview of the Policy Sciences. New York: American Elsevier.

MICHAEL, D. N. (1973) On Learning to Plan–and Planning to Learn. San Francisco: Jossey-Bass.

NYSTRAND, R. O. and L. L. CUNNINGHAM (1974) "Dynamics of local control." Special Report. Washington, D.C.: U.S. Office of Education.

THOMPSON, R. B. (1974) "Projection of enrollments, public schools, city of Detroit, 1975-84." Columbus: Mershon Center, Ohio State University. (mimeo)

WELLINGTON, H. H. and R. K. WINTER, Jr. (1971) The Unions and the Cities. Washington, D.C.: Brookings Institution.

5

Citizen Participation and Urban Water Resource Uses

MICHAEL P. SMITH
DOUGLAS D. ROSE

□ SINCE DIRECT CITIZEN PARTICIPATION in policy-making became an institutionalized component of numerous federally financed intergovernmental programs, much literature has been generated concerning the costs and benefits of citizen participation. This substantial body of literature contains many complex and sometimes mutually contradictory theoretical propositions. In part these contradictions stem from the fact that discussions of citizen participation have proceeded from three distinct theoretical approaches—democratic theory, organizational theory, and public policy analysis—each containing different underlying assumptions and preferred political decision-making models. In part they stem from the paucity of generalizible empirical investigations of direct citizen participation.

Our study is divided into three major parts. The first part organizes the claims of proponents and opponents of direct citizen participation in policy-making into three theoretical perspectives: democratic process, administrative process, and policy consequences. The purposes of this part are to clarify the sometimes murky theoretical propositions on citizen participation contained in much

AUTHORS' NOTE: *Both authors have made co-equal contributions to this chapter.*

of the advocacy literature (and some of the social science literature) and to serve as a necessary theoretical foundation for the empirical portion of our research.

The second part of our study analyzes data on citizen participation in policy-making drawn from one major policy sector, the making of water resources policy in the state of Washington. As part of a larger investigation into the politics of water resources in Washington, John C. Pierce and Harvey R. Doerksen conducted a mail questionnaire survey of the state's heads of households. Pierce has generously sent us the processed data that form the basis of this part of our analysis.[1] The 667 respondents to the Pierce-Doerksen questionnaire gave opinions on a broad range of questions. For our study, responses have been organized to address three central questions: (1) What policy priorities, patterns of citizen participation and representation, and modes of policy-making does the public want? (2) What is the actual pattern of citizen participation in water resource policy development? (3) What results has the citizen participation process yielded?

The third part of our study examines the extent to which various aspects of the theoretical models discussed in the first part of the study were found to obtain for citizen participation in Washington's water use policy. This part concludes by attempting to develop an alternative model of citizen participation in policy-making derived from the preferences, expectations, and perceptions of the sample of citizen participants surveyed in the study.

I. PERSPECTIVES ON CITIZEN PARTICIPATION

Assessments of the costs and benefits of citizen participation depend in part on the vantage point of the observer. Three perspectives in particular are germane to the present analysis. What do proponents and opponents of citizen participation in policy-making expect to be its consequences for democracy, administration, and public policy?

THE DEMOCRATIC PROCESS PERSPECTIVE

Citizen participation in policy-making has support from a variety of political theorists as well as many activists who see it improving the effectiveness of the *democratic process.* At the most general level, direct citizen participation in policy-making is viewed as a good in itself, because it affords the potential for greater democratic responsiveness and accountability by policy-makers. Whatever may be its actual impact in specific cases, some writers argue that *in principle* institutional citizen participation is an important potential channel (in addition to elections, public hearings, referenda, initiative, petitions, and recall) for transmitting popular policy preferences directly to administrators and elected officials. Additionally, through this channel it is sometimes possible for citizens to directly participate in choosing among alternative public policies. Since the potential for channeling public opinion and sharing in public power is always present, citizen participation is valued as an end in itself, regardless of its immediate temporal consequences (Goldblatt, 1968: 35-36).

Several more specific arguments have been raised in support of institutionalized citizen participation in policy-making as a way to hold government more accountable to the public. In summary fashion these arguments go something like this:

1. Institutionalized citizen participation in policy-making channels *policy specific* information concerning citizens' preferences to decision-makers. Such focused, issue-specific input is necessary and desirable because other more broadly aggregative channels for transmitting citizen preferences, such as political parties and elections, are unable to accurately reflect the sectoral policy choices of groups and individual citizens (Stokes, 1963). Through the latter channels, groups and individuals are required to choose people rather than specific policies. Hence such factors as party identification and candidate personality enter the picture as partial determinants of voter choice (Pomper, 1975: chapters 7-9). Even when candidates are chosen because of individual voter or group issue preferences, this choice tends to be either an aggregate "on balance" choice, and hence an imperfect cue to decision-makers concerning policy-specific public preferences, or a choice based on a single intensely felt issue preference, which leaves policy-makers in the dark as to popular sentiment on a whole host of other policy issues. In view of these deficiencies, institutionalized citizen participation is a necessary supplementary device. Appointed citizen participation structures afford the opportunity for the policy-specific representation of a wide variety of directly affected organized and latent community interest groups. Elected citizen participation committees offer voters the chance to choose candidates solely on the basis of their stated

position on the issues at stake in a single policy sector. Hence, citizen participation is a precision instrument for rendering bureaucratic decision-making accountable to either those groups most immediately affected by a policy or to majority sentiment as expressed in policy-specific electoral processes.

2. Once representatives are chosen to sit on citizen participation committees, such bodies will meet regularly with program administrators and expert policy analysts. Program administrators thus will be regularly required to explain and justify their present and future programs to a group of lay citizens, who can call on the advice of experts to test the accuracy of assertions made by administrators. Experts, in turn, will be forced to translate complicated theories and complex quantitative analyses into forms that render their expert knowledge accessible to laymen (as well as to administrators). The lay representatives of both immediately impacted groups and of "everyman" will introduce new information on the likely human costs of policy to both experts and administrators. The entire decision-making process will thus be infused with a degree of mutual awareness and concrete "common sense" that is sometimes lacking when policy is made in either an abstract theoretical vacuum or an insular bureaucratic arena. Greater common sense is simply another way of saying more democratically accountable policy-making.

3. If widely used as an instrument of policy-making, direct citizen participation renders each administrative agency accountable to its own unique set of interests or issue-specific public. This leads to greater accountability *within* each sector. But there is a danger here that each agency may pursue rampantly particularistic policies in pursuit of the needs and interests of its major clientele. According to some democratic theorists (Lindblom, 1959), this danger is minimized because in addition to pursuing its own interests, each agency-clientele policy sector can also serve as a "watchdog" against excessive particularism by every other (or at least its own most directly competitive) policy sector. Thus, the checks and balances provided by pitting policy sector against policy sector also serve to make policy-making *as a whole* more responsible and accountable.

4. Finally, for some democratic theorists citizen participation structures are useful because they are presumed to offer a visible target for the mobilization of heretofore politically unorganized and therefore unrepresented social groups. (For a good critical summary of this position see Piven, 1966.) In this view, such groupings as poor people, ethnic subcommittees, and aggrieved working-class neigh-

borhood residents tend to be underrepresented in traditional channels of political participation. In contrast to such remote traditional structures, citizen participation bodies are located at the point of policy impact where lower-class clientele groups have most of their actual contact with governmental institutions. As such, the citizen participation channel is a visible "target of opportunity" for community organizers seeking to expand the political participation of heretofore politically underrepresented or unrepresented groups. Viewed positively, the mere existence of this channel holds out the promise of formal power-sharing as an incentive for low-income residents or clientele groups to form political coalitions. Viewed negatively, the existence of a citizen participation body that does not presently include representatives of underrepresented constituencies converts the structure into a "target of protest," which can also serve to mobilize the previously unorganized.

Furthermore, to the extent that political mobilization is successful, the now newly mobilized interests are better able to participate in other channels of political action. Citizen participation thus not only improves the viability of the democratic process within particular policy sectors, but also may have a spillover effect on all other political participation channels (e.g., voting, petitions, public hearings). Eventually these structures, too, can be expected to become more democratic as they: represent a wider variety of socioeconomic strata, become more dynamic and participatory, and thereby become more accountable to the publics they serve.

The most comprehensive *criticism* of citizen participation from the perspective of democratic theory is the argument that *in principle,* if democracy is taken to mean equal representation of people (one man one vote), direct citizen participation in policy decisions, below the constitutional level, is *inherently undemocratic.* This is because the one man one vote argument is based on the assumption that all men have an equal stake in politics as a whole. As soon as single-issue areas are considered, however, the presumed equality of stakes among participants disappears. What remains to be represented are a set (or sets) of *interests* or *stakes* rather than *individuals* (Griffiths, 1969: 133-156).

Representing interests or stakes creates a number of difficulties.

1. In terms of cultural values, such representation is likely to be perceived as undemocratic.

2. The basis for legitimate democratic representation, embodied in most current American political institutions, is the representation

of persons not interests. These same structures are not well suited for representing stakes or interests. Thus, if interests or stakeholders were to be effectively represented, new institutions would have to be set up. The establishment of any new set of political institutions historically always has entailed considerable political and economic costs. The establishment of institutions sufficient to accommodate the enormous variety of political interests in a highly complex and heterogeneous society is likely to be especially high. In the American context costs are further increased by the fact that the new institutions would contradict dominant cultural values, which favor the representation of individuals.

3. Interests to be represented vary from policy sector to policy sector and from time to time. This problem of varying interests poses enormous practical difficulties. First is the question: What is the fairest and most appropriate citizen unit to be represented in a given policy sector? For instance, with respect to the issue of community control of schools (Fantini et al., 1970: chapter 8) what is the appropriate "community" of interest that ought to participate in running the school system? Is it all parents in a particular area, who presumably have a direct stake in the quality of their children's education; the students, who are the most immediate consumers of the public service; all neighborhood residents, to gain the broadest base of views; all taxpayers, to insure fiscal integrity and accountability; or the dominant ethnic group in a particular locale, to insure the sensitivity of the school system to particular subcultural life styles? The point to be made is that such practical controversies are inevitable byproducts of institutionalized citizen participation in policy-making.

A second practical problem in representing interests rather than individuals concerns the methods used to select representatives. For instance, *appointed* citizen advisory committees can be expected to be unrepresentative of the actual range of affected interests in a given policy area because they may well be stacked with representatives of groups that administrative officials expect to be supportive of an agency point of view. Similarly, appointive citizen participation in the federal urban renewal program has been criticized because program guidelines call for the institutionalized participation of only "established groups in the community" (Spiegel, 1968: 25), thereby shortchanging latent but presently unorganized community interests such as the low-income residents of areas that are targeted for renewal. Alternatively, *elected* citizen participation boards (as in the example of Community Action and Model Cities elections) are criticized for their unrepresentativeness because of the very low turnout that has often characterized such elections, particularly in low-income neighborhoods (Lowi, 1969: 243). Low voter turnout is believed to favor the most politically organized sectors of the community (e.g., church related groups, organized service providers) and to underrepresent other affected interests.

4. There is no operational principle for disaggregating "one man one vote" within policy sectors (Arrow, 1963). Therefore policy-makers need a substitute basis for determining who has an "interest" in the policy. This generally leads to the use of *objective indicators* of "stake holding" as a guide for the inclusion of interests. The use of such indicators reflects current inequalities in objective conditions. That is, actors with the most resources have the most "stakes" in policy outcomes. As a result, summing up *across* policy sectors, there is an inequality in the representation of individual citizens. For example, the poor have fewer financial stakes (an objective indicator) in all policy decisions than do the rich. "Interest" representation thus does not lead to one man one vote for both rich and poor, but rather to few votes for the poor and many votes for the rich, and is therefore undemocratic. In addition, *within* most policy sectors the greatest "stakeholders" are likely to be organized service user groups and administrators themselves, both of whom share the greatest "stake" in the continued growth and survival of the program. Thus, citizen participation in policy-making based solely on the principle of "primary user" representation is also undemocratic, because it unfairly overrepresents the greatest stakeholders and gives short shrift to both individual wants and other affected interests.

A theoretically feasible way to disaggregate individual interests without relying on objective measures of stakeholding is to ask all people to disaggregate their own perceived stakes into policy sectors. Representation of each individual in each policy area would then be based on the individual's perceived stakes in each area. However, to ask all people to rank order all of their political preferences in this way, so that they might be counted and summed for all policy domains, would be in principle, let alone in practice, impossible.

Since in principle *all* cannot directly participate subjectively, some theorists (Dahl, 1966: 134-151) contend that a political system may be called democratic if at least the most intensely aggrieved interests are allowed to participate in policy-making. Such self-selected citizen participation in policy-making, which nominally uses a subjective estimate (i.e., intensity) of stakes as the basis for representation, in actuality involves selection on the basis of such objective correlates of intensity as political skill, time, energy, awareness, and organizational capacity. This, in turn, leads once again to the unequal representation of persons in policy-making. The upshot of the foregoing analysis is that there is simply no way to obtain direct "citizen" participation in policy-making which is democratic.

5. Accordingly, institutionalized citizen participation that represents interests rather than individuals provides only the illusion but not the reality of democratic political participation. Administrators and "a few big interests"[2] have many political advantages such as expertise, time, and organization. Furthermore, the actual resources under the potential control of citizen participation bodies in policy sectors that encompass small stakeholders (e.g., urban ghetto social services) are rather slight (Altshuler, 1970: 53-56). Hence, formal citizen participation is at worst a sham and at best a dead end.

Despite the fact that direct "citizen participation" in policy-making favors large *stakeholders,* the process nonetheless offers *symbolic representation* to *persons.* This symbolic eyewash of "citizen control" serves to placate many latently aggrieved members of the general public. Such symbolic placation can be expected to lead to generalized political quiescence in the face of official policies which might otherwise engender controversy (Edelman, 1964: 22-43). Institutionalized citizen participation is thus likely to stand in the way of actual protest by individual citizens, which emerges spontaneously in response to felt grievances.

In sum, from this perspective, "citizen participation" is envisaged as a tool of political elites and organized service user groups rather than as a channel for the representation of members of the mass public. It is thus depicted as an undemocratic, if not antidemocratic, instrument, which ought to be resisted.

For some writers, however, interests rather than persons *should* be represented in policy-making (DeGrazia, 1958: 113-122). Citizen participation threatens the proper "democratic" (i.e., interest representation) process for such writers *precisely because* the symbols of citizen representation imply democratic representation of persons and these symbols might be taken seriously by members of the mass public. That is, if community bias is aroused, as in school busing or placement of public housing or fluoridation issues, citizen participation channels may lead to demands that the preferences of a majority of the people predominate. There is no resolution to this situation: if people are represented, major stakes—those of black students or low-income tenants—are not, and bad policy results; if major interests rather than people are represented, then the outside community will be alienated, uncooperative, and generally will threaten the long-term stability of the policy- and decision-making system.

THE ADMINISTRATIVE PROCESS PERSPECTIVE

Institutionalized citizen involvement in policy-making also has advocates who see it benefiting *the administrative process.* In this view, citizen participation is perceived as an opportunity for administrators to mobilize support from organized and latent community groups for the policy agenda of the administrative agency. In confronting other power holders (e.g., executives, city councils, functionally competing agencies, resource suppliers) with alternative proposals, the credibility and legitimacy of agency preferences is enhanced by support from citizen participation structures. The chance to persuade affected citizen groups of the wisdom of agency proposals can thereby provide a useful "bargaining tool" to the agency (Davis, 1973: 61-62).

While citizen participation may legitimate agency preferences for other decision-makers, it can also symbolically legitimate agency policy to the general public. Formal "power-sharing" with citizen representatives reassures the public that general interests have been served, while informal domination of policy-making by administrators keeps actual power in agency hands. Both internal and external legitimation benefits require the co-optation of citizen representatives by the agency. The process of formal citizen participation itself is viewed as promising co-optation because citizen involvement produces identification by citizen representation with the policy-making and administrative process and its outcomes. Once citizen representatives are co-opted, their participation in policy-making symbolically legitimizes the administrative point of view to the public. Hence, unlike the democratic process perspective, from the administrative perspective the eyewash of "citizen control" is a *desirable* consequence of institutionalized citizen participation.

This perception of the citizen participation structure as an essentially passive instrument in the hands of administrators is nicely illustrated by the Department of Housing and Urban Development's program guideline for its urban renewal program, a federal-local program whose citizen participation component has often been criticized for fostering "ritualistic" participation (Sigel, 1967; Dahl, 1962; Smith, 1974b). In reply to the rhetorical question "What is the job of the Citizens Advisory Committee?" HUD's Workable Program for Community Improvement (Spiegel, 1968: 23-28) states:

> The primary functions of the committee and its members are:
>
> (1) to learn about the nature and extent of the deficiencies and the means and methods for remedying them;
>
> (2) to make recommendations for improvement; and
>
> (3) to help inform other citizens and groups as to the need for the improvements and thus develop united community understanding of the need.

A second, less passive administrative view of citizen participation sees such institutions as a way to benefit the administrative process by *increasing the sensitivity of administrators* to citizen perspectives on the human consequences of programs, thus making administrators more effective and responsive (Herbert, 1972; Marini, 1971; Strange, 1972). Such increased sensitivity can come about in four distinct ways. First, through a process of *social learning*, an administrator may *internalize* the beliefs, core values, or cultural perspective of affected citizens. Dynamic socialization processes at work over time as administrators interact with representatives of affected citizens may thus serve to reacculturate administrators, originally socialized into a relatively insular set of professional values and beliefs, into more broadly held beliefs and values shared by the general public or by any of a number of directly affected citizen groups. *Role taking* is a second way by which increased contact between administrators and affected citizens can sensitize administrators to viewpoints not previously considered in policy-making. By role taking is meant a process whereby person X, without altering his or her own self-image, nonetheless evaluates a situation from the vantage point of person Y before arriving at a decision. Unlike social learning, where the administrator permanently internalizes new values, role taking simply requires administrators to temporarily put themselves in citizen representatives' shoes. This temporary shift of focus is likely to enhance the administrators' awareness of the social costs and benefits of policy alternatives to immediately affected groups and/or the general public. For example, the public housing administrator who scales down the size and height of a proposed low-income scattered-site housing development in response to the protest of representatives of impacted neighborhood residents voiced at a public hearing is unlikely to have acted because of inner conversion to a middle-income property owners' point of view. The formal constraints and informal role expectations surrounding his job (as a "builder and manager" of low-income housing) are such that if he does alter his plans at all, he is more likely to do so because exposure to the expressed grievances of others has provided him with the opportunity to put himself temporarily in their shoes, but long enough to recognize that they have an intensely felt and a legitimate concern, which can be addressed by scaling down the project.

The example just provided may be used to illustrate the two other ways by which institutionalized citizen participation may alter administrative agency perspectives. Administrators may alter their plans prior to any actual contact with citizen representatives if they *anticipate* intense or widespread community opposition. Consider, as a case in point, the public housing director whose previous experience with community groups at public hearings, advisory committee meetings, and the like has taught him to expect well-organized and highly vocal community pressure against any new housing development larger than, say, two stories high. Such a person may respond to the community perspective only insofar as he comes to believe that failure to lower his own (or federal housing officials') expectations concerning project size will lead to interminable delay. Thus, he may anticipate the reactions of affected residents, reluctantly embody some of their concerns in his proposal, and thereby modify his policy to diminish the likelihood of citizen protest. Some advocates of institutionalized citizen participation believe that the very fact that formal channels for representing affected interests are present in the administrative planning process makes this "rule of anticipated reactions" (Friedrich, 1941: 589-591) a likely by-product.

A final way by which agency officials may modify their actions requires no modification whatsoever of their attitudes, and no a priori internalization of nonagency perspectives. A bureaucrat may change a decision if new information that he previously lacked changes his decisional premises or his calculations about the probable social or political consequences of the decision. Institutionalized citizen participation is sometimes depicted as a major vehicle which can improve the accuracy of information available to administrators

(Davis, 1973: 64-65; Smith, 1971: 662). Thus, to return to our example, the public housing administrator may have thought that his previous attempts to sell his program, to recognize legitimate affected interests, and/or to anticipate community reactions were fully adequate, but he may later discover that some previously unorganized but immediately impacted interest group (e.g., a new nearby block association seeking playground facilities) has not yet been heard from. Leaders of this group may present him with new evidence (e.g., locations of various playgrounds throughout the city, numbers of unserviced children on the block) that may logically convince him to change his plan. Responsiveness to the rules of logic and evidence has changed the decision without changing the administrator.

Despite the weight of the foregoing arguments, not all administrative theorists and practitioners envisage institutionalized citizen participation in policy-making as an unqualified benefit. Some administrationists who oppose the citizen participation requirements of various intergovernmental programs argue that it can be self-defeating for administrators to provide the opportunity for potentially hostile groups to mobilize (Goldblatt, 1968: 37). Such objections are usually voiced by practitioners who depict any involvement of laymen in policy-making as a troublesome and time-consuming nuisance that diverts professional administrators from their jobs. The objections are usually raised by those who see all delays caused by the "messiness" of the political process as an unwelcome inconvenience, whether such "intrusions" are caused by political party demands, interest group pressures, political protest, or institutionalized citizen participation.

Thus, some administrative opponents of citizen participation believe that the ideally best administrative system is one that puts the professional in charge. Administrative experts are viewed as more objective and politically neutral than other political activists, who are seen as narrow, politically biased special-pleaders. (For a study of some policy implications of this set of perceptions see Smith, 1974a.) It is further believed that a long period of training and indoctrination provides professional administrators with the cognitive skills and the set of professional norms that insure their analytical objectivity and political neutrality.

The upshot of these arguments against citizen participation is that expert public administrators are the best judges of the extent to which various policy alternatives serve the "public interest."

Granting such persons wide discretion thereby improves the administrative process by putting the best qualified judges in charge of the administrative system. This administrative perspective perceives citizen participation structures as lowering the quality of information, the level of skills, the political objectivity and neutrality, and thereby the "fairness" of the administrative process.

THE POLICY CONSEQUENCES PERSPECTIVE

Some analysts of citizen participation in policy-making are concerned primarily with its *policy consequences,* whatever may be its impact on democracy or administration. This perspective focuses on the *nature* and *costs* of public policies developed and implemented by means of institutionalized citizen participation, as well as with the societal *impact* of such policies.

Policy advocates of citizen participation believe that it can serve to avoid long delays in the adoption and implementation of public policy. In this view, citizen participation structures provide a legitimate channel whereby (a) citizen preferences can be heard early in the planning process, (b) centrally affected interests can be represented, and (c) a wide variety of directly and indirectly affected interests can participate in policy deliberation. In the absence of such a viable formal channel, policy advocates foresee interminable delay resulting from protest demonstrations, strikes, boycotts, court battles, and other modes of political conflict (Miller and Rein, 1969). Such delays may produce undesirable policy consequences for two reasons. First, given inflation in the prices of goods and services, protracted conflict runs the risk of raising program costs to the point of diminishing returns. Second, protracted conflict increases the controversiality of a policy, the likelihood of loss of political support for it, and thereby, its chances of long-term survival. Hence, it is believed that the short-term time delays and cost increases resulting from institutionalized citizen participation are worth the price because they are likely to yield long-run political and economic cost savings.

Advocates concerned about the long-term societal impact of public policy further believe that citizen participation improves the quality of information about probable social impact that is introduced into policy planning. The greater amount and wider variety of perspectives on who stands to benefit or suffer from a proposed policy is believed to widen the tunnel vision of the expert. The broadened perspective, in turn, is expected to lead to the modification of public policy in order to exclude likely harmful side effects or at least to compensate sufferers (Cahn and Cahn, 1968: 220-222). Furthermore, the introduction of different class and/or subcultural perspectives into the policy-making process is believed to minimize the likelihood that latent class and/or cultural biases held by upper-middle-class professional policy-makers will find their way into public laws, regulations, and agency programs (Altshuler, 1970). Hence, for policy advocates direct citizen participation in policy-making greatly increases the probability that the full implications of a public policy will be traced through before it is adopted. This, in turn, serves to avoid some of the problems of unanticipated social consequences that emerge when adopted policy is implemented.

The most commonly voiced objection to these arguments is the belief that, in practice, citizen participation unduly prolongs the policy planning process, thereby adding to the dollar costs of a program and running the risk of obsolescent policies (Goldblatt, 1968: 41). The planning process, in this view, is prolonged because the institutionalized citizen participation structure is a highly visible "target" which affords community organizers the incentive to mobilize hostile latent interests, which otherwise would remain dormant. Rather than envisaging community organization as an opportunity to improve information on policy impact, or to build support for a program, some policy analysts believe that community organization is an inherently divisive strategy. As such, providing aid and comfort to community organizers by means of institutionalized citizen participation can only serve to prolong the process, whose policy goals might otherwise be set with less controversy, fewer dollar costs, and a greater likelihood of timely impact.

Related to this set of perceptions is the belief by some policy analysts that institutionalized citizen participation is likely to decrease the objectivity of information that is included in the planning process. Lack of objectivity ultimately produces policies whose objectives are inadequate, ill-defined, vaguely worded, and/or mutually contradictory (Lowi, 1969). Such ill-conceived policies are believed to result from the failure to achieve goal consensus at the planning stage. Lack of consensus, in turn, is a direct outgrowth of the inclusion of mutually contradictory community interest groups in the planning process. If highly narrow and thus inherently divisive interests must be reconciled before policies can be adopted, decision-makers have limited options:

(a) They can water down policy objectives which are intensely resisted by citizen groups. In the long run this serves to weaken whatever impact the policy was intended to achieve.

(b) Alternatively, they can settle on a vaguely worded enunciation of policy objectives, representing the lowest common denominator of community consensus. This simply shifts the actual definition of the policy to the later implementation stage. The operationalization of vaguely worded objectives at this stage almost inevitably alienates some groups and interests that had failed to anticipate the policy would move in the direction(s) its implementers have chosen. Unanticipated policy consequences are thus a built-in byproduct of the broad delegation of public authority (Lowi, 1969).

(c) To avoid overly general or ambiguous policy mandates, decision-makers may choose instead to include but to compartmentalize mutually contradictory objectives in a public policy (e.g., redevelopment of the central business district and neighborhood stability in the now defunct urban renewal program). This too has its obvious policy costs. The *simultaneous* pursuit of mutually contradictory objectives is most likely to lead to either the left hand cancelling out what the right hand is doing, or to the most politically viable component (development of the CBD, in our example) dominating the policy domain at the expense of the weaker component (Wilson, 1963: 242-249). If mutually contradictory objectives are pursued *serially* rather than simultaneously, in the long run the program is likely to lose support from both initial support coalitions. Permanently pleasing no one, the policy is unlikely to survive.

Over against this subjective and conflict-ridden image of the political process is an image of the professional "policy scientist" as a highly trained, theoretically sophisticated, detached calculator of the probable costs and benefits of proposed policy programs. (For a good discussion of rational versus political decision-making in the context of citizen participation see Aleshire, 1972: 434-436.) Some who believe that policy analysis is an exact science fear that the citizen participation process will lead to an undesirable *devaluation* of the professional expert's ability to predict future policy outcomes. This is because citizen participation is most likely to mobilize directly and immediately affected interests, such as primary service users and direct sufferers. The political influence wielded by such groups is likely to result in a devaluation by such groups as well as by the nonexpert political and administrative elites who are their targets of expertly calculated evidence concerning: the *general* social costs, the *long-run* consequences, and the localized but *indirect externalities* of public policy programs. Thus, both because the political process is believed to be biased in favor of short-term beneficiaries and direct sufferers and because professional policy analysts are believed to be free from such biases, citizen participation is regarded by some policy analysts as a harbinger of bad public policy.

Other analysts of long-run policy impact are far less optimistic about the prospects for predicting the future course of public policy, even when the widest latitude is given to professional experts. Nonetheless, even these analysts sometimes express the concern that, as commonly practiced, citizen participation tends to overrepresent immediately affected users of public services—e.g., parents of school-aged children rather than childless couples or general taxpayers on citizen advisory committees for school construction; industrial and recreational users rather than water drinkers in water resource policy. User influence or control of service delivery often leads to the passing along of some program costs to nonusers in the form of increased general taxes or potentially harmful externalities—e.g., pollution, service relocation costs, the foreclosing of some future policy options that might serve to benefit unrepresented or underrepresented interests. Thus citizen participation is here faulted not because it represents diverse community interests, but because it fails to fully include all actually and potentially affected interests on a fair and equal basis.

A final argument raised by policy analysts of citizen participation is that although some policy sectors require large-scale regional, national, or international approaches (e.g., air and water pollution, economic redistribution policies, foreign policy), meaningful citizen participation is not possible on such a large scale. (On the democratic dilemmas posed by large size and scale see Dahl, 1967: 953-969.) In this view, the larger the scope and scale of the citizen control unit, the greater is the likelihood that the participation structure will require large-scale organized activity, if involved lay participants are to be effectively heard. This requirement means that large-scale institutionalized citizen participation is likely to (a) fail to represent all directly and indirectly affected interests within the jurisdiction, (b) overrepresent interests that are well organized at the higher level, (c) underrepresent more locally organized interests, and (d) underrepresent unorganized general interests. In effect these biases are believed to produce policy that serves the needs and interests of the most powerful and well organized user groups (e.g., large corporations, and agricultural polluters in state water pollution control policy) at the expense of the ever elusive public interest. Hence, large scale "citizen participation" is once again believed to accord a symbolic cloak of public legitimacy to the exercise of public power by large scale "private governments" (McConnell, 1967; Lakoff, 1973).

In summing up, policy advocates of citizen participation call attention to its possible long-run benefits: lowering program costs, stabilizing a policy area, insuring widespread support for and thereby the long-term survival of a policy, and diminishing its potentially harmful spillover effects. In contrast, policy opponents of citizen participation fear it will increase program costs, and lead at best to shortsighted, ambiguous, or mutually contradictory policies, and at worse to domination of public policy by powerful organized private interests.

II. CITIZEN PARTICIPATION AND WATER POLICY

In all six of the above views, what the public wants, feels, or believes is an important consideration for assessing citizen participation. Public opinion on citizen participation is not well understood, as some of the contradictions among the views illustrate. Using data drawn from a study by John Pierce and Harvey Doerksen, we present in this part an analysis of public opinion in one state in one policy arena—Washington state water policy. Each of the tables is based on the responses of at least 547 individuals. While other publics and other policies may yield different specific results, the important variables and relationships are likely to be the same, and we correspondingly try to focus on the major patterns.

Public opinion is germane to three types of concerns about citizen participation. First, the public has preferences among policies and among methods of policy-making. The *desirability* of citizen participation rests largely on these preferences, as do the likely ultimate consequences of citizen participation. To outline the main dimensions of the public's preferences, we investigate: (a) the public's preferred policy clientele, (b) the preferred level of decision-making, (c) preferences among alternative uses of water, (d) preferred methods of representation of the public, (e) preferred decision-makers, and (f) perceived effectiveness of different representatives. Taken together, these constitute a model of policy-making and its outcomes; we suggest that conformity to this model determines the success of citizen participation and other methods of policy-making.

Second, the public has experiences in attempting to influence water policy. The *representativeness* of citizen participation can be estimated from these experiences. We investigate the prevalence of citizen participation in a variety of modes, then use these participation patterns to estimate the representativeness of citizen participation. These patterns are then related to the representation of the public through elections, party, and ideology. The representativeness of participation is crucial to both the legitimacy of citizen participation and to the likely outcomes of participation.

Third, we focus on the *effectiveness* of public involvement in producing policies, as perceived by the public. After investigating the degree to which these beliefs are founded on bias or ignorance, we examine the impacts of perceived effectiveness on preferences for modes of policy-making and satisfaction with policy. If citizen participation is to reassure the public that their views are being considered, then it must be (a) effective, (b) mistakenly perceived as effective, or (c) the symbols of participation by themselves must produce satisfaction. The question of the effectiveness of citizen participation is thus crucial to understanding such issues as symbolic reassurance, stable policy arenas, and short- versus long-term policy costs.

MEASUREMENTS AND STATISTICS

Original data came in three levels of measurement: nominal (for region), ordinal (for preferences, evaluations, and identifications), and dichotomous (for water use and policy participation). For the ordinal and dichotomous data, transformations were performed to produce a zero to one scoring interval, with ranks treated as equal intervals. All subsequent measurements are based on the zero to one scores. For ranked data, a score of one describes the highest rank, while zero describes the lowest rank. For dichtomous data, a score of one describes participation or use, while a score of zero describes nonparticipation or nonuse.

Four statistics based on the transformed data were used. Means are simple statistical averages of all individuals' scores. Regression coefficients of one variable on another derive from standard two-variable regression analysis. Gini indices describe what share of some total amount (e.g., of a variable) is unequally distributed among individuals. The misrepresentation index measures the difference between the participants' preferences and characteristics and those of the population as a whole. For all these statistics, one would be the normal maximum score. A mean of zero would indicate that no one in the sample used, participated because of, or preferred the object. A regression coefficient of zero would indicate no relationship between two variables. A Gini index of zero would indicate complete equality. A misrepresentation index of zero would describe perfect representation. Both misrepresentation and regression coefficients can take on negative scores up to −1. A negative regression coefficient indicates that more of the independent variable produces less of the dependent variable. A negative misrepresentation describes an underrepresentation of the trait.

URBAN AREAS AND WATER RESOURCES

Nominally, Washington shares with such states as New York, Michigan, or Louisiana the potential for intrastate urban-rural conflicts over water resources: the state has a single large city (Seattle) whose metropolitan area includes over 40% of the population. Seattle's economy is still significantly dependent on port facilities. It is geographically far removed from in-state citizens whose economy depends more on rivers, dams, and the irrigation of farms than it does on business and transportation uses of water. The potential for the development of a distinctive urban interest in water use is mitigated unusually in Washington state by two situations: (1) Washington's mammoth rivers run through the state from rural areas through 12 large dams generating power to urban centers, producing a statewide water system with high levels of interdependence usually associated with intraurban, not intrastate, water systems; (2) politics in Washington is state politics, long dominated by partisan conflict, once the highest in the nation (Francis, 1967: Table III-8), which does not pit region against region. It is also dominated by pressure group conflict (Francis, 1967), with the New Deal type of Democratic coalition cross-cutting regional loyalties and eventually producing a statewide political culture—values, attitudes, identifications, and beliefs basically similar from urban center to rural enclave. Thus, urban areas are not distinct units or actors in Washington's water politics to anywhere near the degree that they are in many other states; instead, the entire state more nearly represents an urban area than it does the typical American state.

THE POLITICAL CULTURE OF WATER RESOURCES

The starting point for Washington's views on water policy is strong public preference (more than five-eighths of the respondents) for seeing the "clientele" of Washington's water resources as a statewide public rather than a national, regional, or local public. This preference defines the realm of legitimate claimants and concerns and specifies the relevant level of decision: three-fourths of the respondents prefer decisions on water policy to be made at the state, not national or local, level. The preferred clientele and preferred level of decision are closely related (gamma = .66, beta = .48); the public wants all legitimate interests to be heard and this requires statewide decision-making.

The processes of representing claimants and of making decisions are distinct roles within policy-making. For Washington's water resources, respondents see the "public as a whole" as the primary claimant and thus as an important legitimate participant in policy-making. Two-thirds of the sample deem it "very important" that "the public be actively involved in making decisions about the use of Washington State's water resources." The public's primary task is to decide what policy priorities are *desirable* overall. As Table 1 indicates, the sample is largely in consensus on the overall priorities among alternative uses of water. With exceptions to be noted below, the public's conception of itself as a cohesive unit is accurate for these priorities: the self-interest of users, social and political identifications, and differences in values have little impact on water use priorities (Rose, 1976: Tables 10-12).

The public's preferences, then, need a mode of expression which (1) can convey consensus and (2) is specific to water policy. As Table 2 shows, the preferred *ways* for the public to influence water policy are largely collective (as opposed to individual) expressions of preferences specific to water policy: citizen advisory groups, public hearings, and petitions. Modes which are neither policy specific nor consensual, such as interest groups and parties, are least preferred for representation. However, on *dissensual* questions—such as the relative

TABLE 1
PUBLIC PREFERENCES: MOST IMPORTANT USES OF WATER

Agriculture	Domestic	Energy	Industry	Preservation	Transportation	Recreation
.78	.78	.67	.40	.40	.24	.23

Entries are the average preference for each use.

TABLE 2
PUBLIC PREFERENCES: WAYS FOR THE GENERAL PUBLIC TO INFLUENCE WATER POLICY

Citizen Advisors	Public Hearings	Petitions and Initia-tives	Elec-tions	State Legis-lature	Govern-ment Agency	Interest Groups	Political Parties
.69	.67	.65	.58	.53	.44	.38	.10

Entries are the average preference for each mode of influence.

priorities of industrial and preservation uses of water or of transportation and recreation uses—party, ideology, identifications, and user groups *are* related to preferences and do perform representation functions in resolving conflicting preferences. These dissensual representational mechanisms also can serve to interrelate policies across issue areas (e.g., when use of water to produce energy affects other energy-related issues). What the public preferences reflect, however, is that dissensus on water use priorities is the exception rather than the rule and so is at best a secondary consideration in representing the public.

A desire for representation of the public in policy-making to decide what is desirable is matched by support for participation of experts to decide what is *possible.* Table 3 indicates that questions of possibility are considered so important that the public overall is more willing to let experts decide policy than to let anyone else decide policy. The table also implies that the public conceives of the most difficult aspect of policy-making: the merging of relatively consensual priorities with relatively consensual possibilities. While experts are given primacy in merging the desirable with the possible,

TABLE 3
PUBLIC PREFERENCES FOR WATER POLICY DECISION-MAKERS AND IMPACT OF PREFERRED WATER CLIENTELE ON PREFERRED DECISION-MAKER

	Decision-Maker						
	Experts	Adminis-trators	Legis-lature	Parties	Groups	Public	Citizen Advisors
Overall score	.72	.63	.52	.08	.37	.58	.64
Rank if prefer:							
National clientele	1	2	3	7	6	5	4
Local clientele	2	4	5	7	4	3	1

Entries are the average preference for each decision-maker.

citizen advisory groups are not far behind. The results generally show a preference for *policy-specific* merging of desirabilities and possibilities, as decision-makers more traditionally concerned with interrelating possibilities (administration), interrelating desirabilities (parties and groups), or merging desirabilities and possibilities across many policies (legislature) are less preferred decision-makers than the policy-specific agents. As Table 3 also indicates, the public generally relates their preferences for decision-making modes to their water clientele preferences. If the clientele is defined as local citizens, fairly direct public representatives are the preferred decision-makers, but as the clientele is more broadly conceived as regional or national, decisions are increasingly to be turned over to experts and administrators, with an enlarged role for "preference aggregating" devices such as the legislature. This appears a realistic relationship in that possibilities are more complex for broader clienteles and because preferences are difficult to register for broader clienteles.

As stressed so far, the questions of representation and decision-making are not entirely separate. In terms of the desired *level* of decision and the decision-making *agency,* representation is a key consideration. Without the perceived need for formal representation of the public, experts and administrators are the first choices for decision-makers, with parties, interest groups and the general public itself seen as having little to contribute (Table 4). However, the preference for formal representation vastly increases the decision-making role assigned to the public, with corresponding decreases in preferences for experts, administrators, and the legislature as decision-makers. When the preference for public involvement is strong, the sample has nearly equal preferences for experts, citizen advisory groups, the public itself, administrators, and the legislature as decision-makers, suggesting that no single decision-maker com-

TABLE 4
PUBLIC PREFERENCES FOR DECISION-MAKERS, BY PREFERENCE FOR PUBLIC INVOLVEMENT

	Experts	Administrators	Legislature	Citizen Advisors	Public	Groups	Parties
Public involvement important:							
No	.85[a]	.72	.60	.57	.33	.34	.13
Yes	.69	.61	.59	.65	.63	.37	.07

a. Entry is the average preference for experts as decision-makers among respondents regarding public involvement in decision-making as unimportant.

TABLE 5
PUBLIC PREFERENCES FOR DECISION-MAKERS, BY PREFERENCE FOR REPRESENTATIVE

	Experts	Adminis-trators	Legis-lature	Parties	Groups	Public	Citizen Advisors
Decision-Maker preferred for representative:							
No	.73[a]	.63	.35	.04	.22	.43	.37
Yes	.71	.64	.68	.42	.62	.69	.76

a. Entry is the average preference for experts as decision-makers among respondents not preferring experts as representatives.

bines all the required policy-making roles and that some multistage, multi-decision-maker arrangement is most preferred. Table 5 demonstrates that all agents except experts and administrators have their decision-making justifications largely grounded in their representational function. Thus, the public's preferences for decision-making lie in the combination of expert advice on possibilities and representation of the public's views of policy desirability in specific policy areas. No single agent combines these traits very well in the public's estimate, and the question of representation is particularly open.

PATTERNS OF PARTICIPATION AND REPRESENTATION IN WATER POLICY-MAKING

The public's preferences for water policy-making apparently could be realized. There is a statewide water clientele defined both by the actual patterns of water usage, which link virtually all parts of the state, and by the consensual public preferences for policy priorities. Beyond these requirements, fulfillment of the public's preferences

TABLE 6
PUBLIC PARTICIPATION: "HAVE YOU EVER ATTEMPTED TO INFLUENCE A DECISION ABOUT THE USE OF WATER THROUGH ANY OF THE FOLLOWING METHODS?" (in percentages[a])

Petition	Hearing	Legislature	Agency	Group	Citizen Advisors
34	19	10	9	7	3

Only One Method	Two	Three	Four	Five or Six	One or More Methods
24	11	7	3	1	45

a. Entries are the percentage of respondents with the appropriate category or number of "yes" responses.

TABLE 7
MISREPRESENTATION OF WATER USE PREFERENCES THROUGH METHODS OF PARTICIPATION

	Participation Method							
Use Preference	**Petition**	**Hearing**	**Legislature**	**Agency**	**Group**	**Citizen Advisors**	**Average**	**Any**
Agriculture	–.02	.00	–.02	.01	–.12	–.06	–.03	–.01
Domestic	–.02	.01	–.03	–.04	–.05	–.01	–.04	–.02
Energy	–.02	–.05	–.04	–.04	–.05	–.07	–.04	–.02
Industry	–.03	.01	.00	.03	–.02	.02	.00	–.02
Preservation	.04	.02	.00	.00	.07	.04	.03	.03
Transportation	–.01	–.03	–.02	–.04	.00	.03	–.01	–.02
Recreation	.05	.04	.07	.08	.11	.07	.07	.04

Entries are the score for the misrepresentation of the preference (row) by the participation method (column).

for strong citizen influence on water policy requires (1) a large amount of citizen participation, (2) participation which is representative of public preferences, and (3) representative participation, which is (a) effectively channelled to policy-makers and (b) effective in producing the desired policy. If the public will not or cannot represent its views, or if policy-makers do not act on those views, the respondents' desired model of water policy-making will not be realized. These three requirements are now considered in turn.

1. The Degree of Participation. Citizen participation attempting to influence Washington water policy is both widespread and various. Forty-five percent of the sample has directly participated in water policy-making through one or more of six methods.[3] Such widespread participation in a single, relatively narrow policy sector is surprising, as it exceeds public voting participation in the 1974 congressional elections. The participation is spread widely among citizens; there are a large set of nominal participators and a small set of intense participators, as has been found in the past (Clausen et al., 1965: Figure 1). Multimethod participation is widespread according to Table 6, yielding an equality of participation (Gini index = .68) more characteristic of congressional (Gini index = .59) and presidential (Gini index = .45) elections than of interest representation (Gini index = .94; Clausen et al., 1965). Table 6 also indicates that the participation is spread widely among different methods of participation. The amounts of participation in each method appear roughly proportional to the single-use capacity of the method: that is, citizen advisory committees are small, so a small minority of citizens actually serve on them, while petitions can accommodate many signatures, so a large minority of citizens actually sign petitions. In sum, in very large numbers, citizens attempt to make their preferences clear to decision-makers in a variety of ways.

2. The Representativeness of Channels. The public's model might still be dismissed because participants could be unrepresentative of the general public. Table 7 shows the amount of misrepresentation of the public's policy priorities for the whole set of participators and for participants in each of the six participation channels.[4] Most striking is the overall pattern of representative participation. Energy preferences are consistently underrepresented, but the magnitude of the underrepresentation is small. Only the recreation preferences are consistently overrepresented. None of the channels involve reordering of priorities except when the priorities are virtually tied in the overall preference hierarchy of the general public. Participants in any of the channels provide a fairly accurate sample of the public's preferences. Taken together, the different modes of participation provide a highly accurate picture of public preferences toward water use priorities in Washington.

An additional consideration of representation concerns the relationship between participation and preferences in the specific issue sector of water policy and the participation and preferences that involve all issues: ideology, political party, and voting. Ideology is a way of unifying preferences on all issues, party is a way of forming coalitions across all issues, and voting is a way of participating in choosing decision-makers for all issues. Without issue-specific citizen participation, popular control over water policy would depend on the general-purpose vehicles of ideology, party, and elections. To what degree does the issue-specific participation change the nature of representation? Liberals and conservatives, Democrats and Republicans, and voters and nonvoters have the same overall preferences, and the general hierarchy of preferences is the same for all these groups as well as for issue-specific participants and nonparticipants. As Table 8 shows, however, there are links between water use preferences and the across-issue organizing devices of ideology, party, and voting participation. Strong conservatives are less in favor of preservation and recreation, more in favor of industrial, energy, and domestic uses of water than is the population as a whole. Strong liberals show the opposite differences from the population. Thus, on the most controversial aspects of water policy, preferences are linked to ideology. Similarly, for political party, Democrats are more in favor of recreation and preservation, less in favor of energy and industrial uses of water than is the population as a whole. Republicans show equivalent and opposite tendencies. Though the

TABLE 8
MISREPRESENTATION OF WATER USE PREFERENCES BY PARTY IDENTIFICATION, IDEOLOGICAL IDENTIFICATION, AND VOTERS

Representation Group	Agriculture	Domestic	Energy	Industry	Preservation	Transportation	Recreation
Democrats	.02	–.02	–.05	–.03	.03	.00	.05
Liberals	–.02	–.05	–.04	–.06	.11	–.01	.07
Voters	.00	–.01	.01	.03	–.03	.00	–.01

Entries are the score for misrepresentation of the preference (column) by the representative group (row).

party differences are not as large as the ideological differences, the same general pattern of division on the controversial aspects of water policy exists. Party and ideology would seem to be adequate vehicles for handling the major water use controversies through the general-purpose vehicles of elections and legislatures. Just as in the case of participants in issue-specific channels, party and ideology also carry consensus, and the only preferences consistently over- or under-represented among issue-specific participants are the preferences that consistently split along party and ideological lines—the major disagreements on policy priorities. The magnitudes of the misrepresentation in the issue-specific participation and in the general-purpose vehicles also correspond. Not surprisingly, given the general redundancy so far, election participants carry about the same misrepresentation of policy preferences as do participants in the water policy channels. Table 9 demonstrates that one of the reasons for the redundancy is that voters tend to be water policy participants and nonvoters tend to be nonparticipants in all channels, though most notably in the least "public" channels. Party and ideology, on the other hand, bear little relationship to water policy participation except for mildly underrepresenting liberals and Democrats in citizen advisory groups. The general picture is one of redundancy: issue-specific channels involve the same types of individuals and the same types of preferences as do the more general participation modes. Further, to the degree that policy preferences on water uses are dissensual, the general vehicles of party and ideology tend to organize the opposing points of view. The implication for citizen participation on specific policy is clear: the rationale for citizen participation lies not in the nature of the views represented but in *how* and *where* they are represented. This, in turn, reflects the public's criteria for representation and decision-making presented in the last section: any method that can apply the public's views to

TABLE 9
MISREPRESENTATION OF DEMOCRATS, LIBERALS, AND NONVOTERS BY METHODS OF PARTICIPATION

	Petition	Hearing	Legislature	Agency	Group	Citizen Advisors	Average
Democrats	.00	.00	–.04	–.03	.05	–.06	–.01
Liberals	.01	–.02	–.04	.00	.01	–.06	–.02
Nonvoters	–.04	–.08	–.10	–.10	–.05	–.13	–.08

Entries are the score for misrepresentation of the method of participation (column) by the representative group (row).

policy-making is welcome; the difficulty—and the argument for policy-specific participation—lies in applying the views to specific policies, not in representing the public.

3. The Effectiveness of Public Control of Water Policy. The public's views are generally well represented in all issue-specific and general-purpose channels. The public's policy-making model calls for this participation and representation to influence policy, but it is at this stage that the public's model breaks down. Despite the great desire for public influence and public participation that represents public preferences, the sample feels that the public does not have much influence on Washington's actual water policy. Only 3% of the respondents described the public's influence as "great," only 27% described it as "some," while 58% answered "not much," and 12% said "none." This does not appear to be a biased perception. Participants and nonparticipants gave the same answers, so neither "ignorance" nor "intensity" explains the results.

One clue to public dissatisfaction is given in Table 10. Individuals who have contacted agencies are less likely than those who have not to want decision-making left in the hands of administrators, while individuals who have served on citizen advisory boards, signed petitions, or contacted the legislature are more willing than are the inexperienced to yield decision-making to these agents. This suggests that relatively direct citizen control may be required for policy satisfaction.

The perceived effectiveness of a participation method's representation strongly influences (average b = .16) the preference for using that method for representation. Are these perceptions accurate? In Table 11 we indicate the impact of participation in a channel on the estimate of the effectiveness of *that* channel in providing public

TABLE 10
IMPACT OF PARTICIPATION ON PREFERRED DECISION-MAKER

Type of Participation:	Citizen Advisory Group	Petition	Legislature	Hearing	Agency	Group
Preferred Decision-Maker	Citizen Advisory Group	Public	Legislature	Public	Administrator	Group
Impact of Participation:	.13[a]	.07	.04	.01	−.07	.00

a. Entry is the regression coefficient of preferred decision-maker on the specified type of participation.

TABLE 11
IMPACT OF PARTICIPATION ON
PERCEIVED EFFECTIVENESS AS REPRESENTATIVE OF PUBLIC

Type of Respondent Participation:	Citizen Advisory Group	Petition	Legislature	Hearing	Agency	Group
Perceived Effectiveness of Mode as Representative of Public:	Citizen Advisory Group	Petition	Legislature	Agency	Agency	Group
Impact of Participation:	.00[a]	.00	–.04	–.03	–.05	–.03

a. Entry is the regression coefficient of perceived effectiveness on the type of participation.

influence on policy. Only petitions seem to convince participants of their effectiveness. Experience with agencies, legislatures, hearings, and interest groups convinces individuals that these modes are inadequate. As Table 12 shows, much of the perception of agencies as effective representatives comes from the *least* experienced individuals, those who do not vote. Moreover, experience with the general vehicle of elections convinces voters that issue-specific referenda and petitions are important modes for public influence. These results indicate that the public's perceptions of policy-making, which cannot or do not take citizen preferences seriously, are largely accurate.

For the minority that sees public influence as adequate, the general-purpose methods of elections and the legislature both are seen as the most effective representatives of the public and show the greatest differences between those perceiving effective and ineffective public influence across all channels (Table 13). The majority

TABLE 12
IMPACT OF ELECTION PARTICIPATION ON
PERCEIVED EFFECTIVENESS AS PUBLIC REPRESENTATIVE

Perceived Effectiveness of Mode as Representative of Public:	Effectiveness for: Nonvoters	Voters
Petition	.52[a]	.65
Election	.60	.58
Citizen advisory group	.58	.57
Group	.51	.53
Party	.22	.24
Legislature	.61	.63
Agency	.47	.32

a. Entry is the average perceived effectiveness of petitions among nonvoters.

TABLE 13
IMPACT OF PERCEIVED PUBLIC EFFECTIVENESS ON PERCEIVED EFFECTIVENESS OF MODES OF REPRESENTATION

	Public Seen As:	
Representative:	Ineffective	Effective
Petition	.65	.58[a]
Election	.54	.66
Citizen advisory group	.58	.56
Group	.55	.48
Party	.25	.21
Legislature	.57	.71
Agency	.38	.31

a. Entry is the average perceived effectiveness of petitions among those who see the public as ineffective.

that perceives an ineffective public shows a much stronger preference for direct, unmanaged channels for public representation—elections, petitions, groups—than do those who see the public as already effectively represented (Table 14). Accordingly, neither group sees the issue-specific channels as carrying the burden of public influence. Those seeing an already effective public are willing to leave decisions to experts, administrators, and the legislature, whereas the dissatisfied majority wants decision-making given to citizen advisory groups, the public itself, and interest groups.

TABLE 14
IMPACT OF PERCEIVED PUBLIC INFLUENCE ON PREFERRED REPRESENTATIVE AND DECISION-MAKER

	Preferred Representative if Public Influence is Perceived as:		Preferred Decision-Maker if Public Influence is Perceived as:	
Representative or Decision-Maker:	Low	High	Low	High
Public			.62[a]	.53
Election	.63	.50		
Petition	.69	.61		
Hearing	.64	.71		
Citizen advisory group	.66	.73	.68	.58
Group	.42	.31	.40	.31
Party	.09	.11	.07	.09
Legislature	.50	.57	.48	.58
Agency	.41	.49	.61	.66
Experts			.67	.77

a. Entry is the average preference for the public as decision-maker among respondents perceiving low public influence.

The pattern of actual policy-making is: (1) public views on water policy, while well-represented in general vehicles of elections, parties, and ideology, do not effectively exert public influence; (2) issue-specific vehicles, supposed to directly apply preferences to water policy, again carry the views but do not change policy unless decision-making is largely in the hands of the public; and (3) a disgruntled majority, perceiving both a skewing of representation and a lack of public impact when bureaucrats or elected officials are in control, is more willing to circumvent the established representation and decision-making processes to establish some approximation to direct democracy and direct citizen control over specific decisions.

A major problem with most modes of representation and influence is that they are oriented either to conflicts or to highly specific programs, while what the public desires to impose as a control is mostly an overall consensus on major alternatives. As a result, judgments of effectiveness are only likely to vary when individuals vary on the sorts of issues to which institutions are oriented. That is, as indicated in Table 15, the adherents of the party controlling the legislature and administrative agencies (Democrats) are somewhat more likely to see these as effective than are adherents of the out party—who prefer elections, and petitions, their methods of changing and bypassing party control—because party does organize the general organs of government. Similarly, as Table 16 shows, proponents of water uses already institutionalized into the representation system —such as business uses—see these modes as more effective than proponents of the avocational uses that are not yet institutionalized. The latter prefer methods that bypass the general organs. These differences are not dramatic because only part of what people care

TABLE 15
PERCEIVED EFFECTIVENESS AS PUBLIC REPRESENTATIVE, BY PARTY AND IDEOLOGY

	Difference of	
Representative:	**Party**	**Ideology**
Election	–.14[a]	–.07
Petition	–.12	–.02
Citizen advisory group	–.01	–.01
Group	–.06	.02
Party	.00	–.03
Legislature	.11	–.04
Agency	.20	.12

a. Entry is the regression coefficient of perceived effectiveness on party identification.

TABLE 16
IMPACT OF POLICY PREFERENCE ON
PERCEIVED EFFECTIVENESS AS PUBLIC REPRESENTATIVE

	Agri-culture	Domestic	Energy	Industry	Preser-vation	Recre-ation	Trans-portation
Election	–.09	–.08	.05	–.03	.01	.06	.02
Petition	.01	–.07	–.02	–.15	.06	.12	.04
Citizen advisory group	.00	–.01	.03	–.07	.05	.01	.01
Group	–.05	.08	–.05	–.03	.03	.03	–.04
Party	–.01	.09	–.08	.14	–.07	.01	.00
Legislature	.08	.03	.05	.15	–.11	–.15	.04
Agency	.05	–.03	.03	.02	.07	–.08	–.05

Entries are the regression coefficients of representative's effectiveness (row) on preference for water use (column).

about is organized into policy-making. For the most part, the unrepresented consensus leads almost everyone to view citizen control over policy as grossly inadequate.

CITIZEN CONTROL OVER POLICY AND POLICY SATISFACTION

Given the difficulties of citizen participation, one strategy that policy implementers might follow would be to ignore the citizen input and concentrate on providing programs that are needed and wanted, regardless of how these programs are decided upon. In part, this is "government as usual," and it has yielded policies that neither satisfy nor dissatisfy the public (mean policy satisfaction among public = .56). Simply by changing policies, there does not appear to be much likelihood of increased satisfaction. As Table 17 shows, individuals who favor water uses already supported by public opinion and public policy are somewhat more satisfied than individuals preferring less consensual uses that require new policies; but satisfaction is not very readily explained by these results–the

TABLE 17
IMPACT OF POLICY PREFERENCE ON SATISFACTION
WITH WATER POLICY

	Agri-culture	Domestic	Energy	Industry	Preser-vation	Recre-ation	Trans-portation
Impact of use preference	a	.08[b]	.08	.09	–.11	–.16	a

a. Insignificant because of too little variance in use preference.
b. Entries are regression coefficients of satisfaction on preference for use.

TABLE 18
IMPACT OF POLICY SATISFACTION ON
PERCEIVED EFFECTIVENESS AS PUBLIC REPRESENTATIVE

Effectiveness Rank Among:	Legis-lature	Election	Petition	Citizen Advisory Group	Group	Agency	Party
Satisfied	1	2	3	4	5	6	7
Dissatisfied	5	4	1	2	3	6	7

differences between proponents of alternative priorities are not that large, and a change of priorities would, except in the case of industrial and preservation priorities, result in increased dissatisfaction.

In contrast to this pattern, the impact of *estimates of public influence* on policy satisfaction is .27, two or three times the impact of policy preferences. Individuals who perceive citizen control over water policy are 60% more satisfied with policy than citizens who do not perceive citizen control. If the public is to be satisfied with policy, they probably will have to see citizen participation as a major control over policy.

Government as usual already includes some citizen control over policy. The textbook picture of policy-making is that the public is represented through elections and elected officials; then the elected officials, administrative officials, and experts make policy by combining what the public wants with what is feasible. Tables 18 and 19 illustrate that this textbook picture is accepted as adequate by individuals satisfied with policy. The textbook solution to dissatisfaction is to vote the rascals out, but that is not the solution those dissatisfied with water policy are now advocating. The dissatisfied see the policy-specific methods of petitions, citizen advisory committees, and interest groups as the effective public representatives, and they prefer abandoning the textbook system in favor of citizen advisory committees and direct democracy as ways of deciding policy. Policy dissatisfaction based on lack of public control over policy provides

TABLE 19
IMPACT OF POLICY SATISFACTION ON PREFERRED DECISION-MAKER

Preference Rank Among:	Expert	Adminis-trator	Legis-lature	Citizen Advisory Group	Public	Group	Party
Satisfied	1	2	3	4	5	6	7
Dissatisfied	3	4	6	1	2	5	7

sizable public support for replacing general public representatives and general decision-makers with direct public control over specific policies. This support exists despite the lack of a single method of public control that the public perceives as an adequate method of policy-making. If such a method were found, support for making issue-specific citizen participation the basic mode of policy-making would probably be quite widespread.

III. CONCLUSIONS AND IMPLICATIONS

Both the arguments and the evidence about citizen participation start from preferred models of the policy process. These models in turn influence behavior and its interpretation and pose practical problems with citizen participation.

PUBLIC PREFERENCES FOR CITIZEN PARTICIPATION

The model of public policy explicated by public opinion starts from the premise that water policy should benefit a statewide mass public and hence should be decided at the statewide level. This preferred large scale of decision-making poses structural representation problems, as argued by opponents of citizen participation from the democratic and administrative processes perspectives. It also appears to bar some of the indirect community organizing benefits contemplated by some advocates of citizen participation.

Public opinion further prefers that the *general* public be influential in policy-making. This preference accords both with the notion that citizen participation is good in itself and with the notion that citizen participation can legitimate policy processes and outcomes. It runs counter to the wish of some opponents of citizen participation to keep "subjective" information out of the policy process: the public believes that its "subjective" preferences are a legitimate determinant of public policy. But the public prefers no tyranny of the majority. Its preferred modes of representation—citizen advisory groups, hearings, and petitions—are policy-specific rather than generalized governing mechanisms. As argued by democratic proponents of citizen participation, the great weakness of public influence at present is the limited number of modes for expressing individual citizen's preferences on specific programmatic choices; just as opponents point out, this limitation requires new decision-making structures and different structures for each policy. However, contrary to the view of these structures as unrepresentative, the public clearly perceives citizen advisory groups and hearings as democratic mechanisms. Since two of the four preferred representation modes —petitions and elections—clearly count *people,* not interests, and since the other two favored modes—hearings and citizen advisory groups—do not appear in the public's view to be interest-representing structures, the public does not seem to prefer interest-representing citizen participation methods.

A final element of public preferences toward the policy-making process concerns the choice of decision-makers. Here the public's preferences are not arranged along a simple public versus agency dimension as contemplated in many of the arguments for and against citizen participation. The public prefers both administrators and experts on the one hand and citizen advisory groups and the general public on the other hand as decision-making participants. Rather clearly, the public values professional, "objective" input about *possibilities* and democratic, "subjective" input about *desirabilities.* It does not see the two as necessarily incompatible, as do policy analysts who oppose citizen participation. The public is less concerned with calling administrators per se to account, as expected in the democratic theory justification for citizen participation, than with informing or educating administrators as expected by those who wish citizen participation to improve the administrative process. As opponents of citizen participation expect, the ensuing lack of simple agency dominance does make the policy process messier, but, by downplaying mediating institutions, it also makes the process much more straightforward for particular policies.

The public's preferred policy-making model thus fully fits none of the theoretical perspectives and contains some elements not clearly elucidated in any of them. Much of the concern about divisions among the public shared by many analysts is simply not shared by the public itself; citizen participation in specific policies flows from a first principle among the public that policy should be dominated by and directly benefit the *general* public, a principle given, at best, lip service in many of the perspectives. Much of the concern with interrelating interests, policies, and programs is again not shared by the public; nor are the problems of aggregating individual preferences a major concern of the public. The public, unlike most of the theories, contemplates a variety of direct citizen participation modes as both desirable and possible. Finally, the public is radically willing to exclude from policy-making such mediating institutions as parties, interest groups, and perhaps even legislatures and executive coordinating structures.

PATTERNS OF CITIZEN PARTICIPATION IN WATER POLICY

Thus far, we have summarized only what the public prefers. We now turn to the correspondence of actions and attitudes to the public's preferred model. First, the public's conception of itself as a relatively unified clientele seems quite in keeping with its generally consensual policy preferences for water use. Thus some of the opposition to citizen participation by policy analysts, based on the assumption that no consensus exists for a statewide public, is inappropriate in this case and perhaps in many other cases.

Second, the public's preference for broad public involvement in specific decisions is matched by broad participation in water policy-making, supporting the argument that democratic participation in specific policies is possible. Although, as policy analysts who oppose citizen participation expect, active participants are harder to satisfy than nonparticipants, the difference is not due to their policy-specific participation: the largest differences in terms of demands are between voters and nonvoters, not between persons who only vote and those who attempt to influence specific water policy decisions. Additionally, the participation is thus broadly representative of public preferences, thus improving the information available to decision-makers, as argued from the policy consequences standpoint. Furthermore, because participation in one mode does not tend to produce participation in other modes, the use of one policy grievance to stimulate further organization, one of the expected political consequences of citizen participation, does not appear to apply in this policy arena.

Third, the public prefers a variety of modes which express individual preferences for specific policies. This preference corresponds to the actual patterns of participation and representation. Direct expression modes such as petitions and hearings are most widely used, whereas mediated modes–such as interest groups, administrative agency contacts, and the legislature–occur less frequently. In addition to direct expression, use follows the capacity of a method to express a multitude of individuals' preferences: petitions are most used and citizen advisory groups involve only a small number of individuals. All six channels receive relatively representative inputs, so discussions of "captive channels," representation of interests not people, selective modes, skill and intensity bias, and the like seem of secondary importance. Nonetheless, one channel, the citizen advisory committee, does seem to represent interests, particularly those of business users, conservatives, Republicans, and intense transportation users, more than do the other channels. The narrower low-participation channels are somewhat more likely to be influenced by more skilled and organized citizens. On the other hand, these sorts of considerations appear to be no greater a problem for citizen participation in policy than for elections and are likely to be considerably less severe than in policy processes not involving any citizen participation. Perhaps a greater problem exists with using citizen participation channels to register preferences without expressly requiring participants to register a full preference ordering: intensity and misrepresentation occur most often with regard to relatively *low-priority* water uses even for users, such as recreation. Without an opportunity to consider a full set of preferences, considerable support for low-priority uses might be registered.

Fourth, the public's preference for a strong hand in actual decision-making has not been realized in practice. Participants are not particularly satisfied with policy, policy-making, or the public's influence, and a number of arguments contemplating a satisfied public following from citizen participation do not apply. The participation does not appear to have legitimized policy, nor do administrators appear to have been sensitized by citizen participation. In fact, agencies in particular are viewed as unresponsive by participants. The voting public started out hostile and alienated and did not change much. Arguments that citizen participation will bring greater participant acceptance of policy appear to apply only when participation is *formal, visible,* and *influential.* Only the citizen advisory groups and petitions seem even to promise the prospects of these sorts of changes, and even these are conditional on policy that does what participants want. Several recommendations emerge from these findings:

(1) citizen participation may be very useful to give primacy to broad, consensual preferences that might otherwise be overlooked or contravened by other actors, as such consensual preferences do not follow district, party, ideological, self-interest, or other traditional lines of political organization;

(2) administrators can expect direct participation to be serious, widespread and relatively representative;

(3) it appears that special procedures to guarantee representativeness —such as stratified representation of users or discounting the value of "one-sided" petitions—are as likely to produce misrepresentation as to correct it, so such procedures should be used only when the population's *basic* policy preferences are already known;

(4) because all the modes of participation considered here involve relatively representative input, none should be discounted and the choices among them can more appropriately be made on procedural rather than substantive grounds;

(5) citizen participation by itself appears to be most potentially useful for producing policy acceptable to the general public, but this potential will probably remain unrealized unless citizen participation is accompanied by a number of other changes in the policy process designed to insure citizen effectiveness.

THE IMPLICATIONS OF CITIZEN INEFFICACY IN POLICY-MAKING

The problem of citizen ineffectiveness is, for the general public, the most serious procedural problem in the policy-making process. As citizen participation is often conceived of as a solution to this problem, a review of the failure of citizen participation in the case of water policy suggests some fruitful changes. The key problem is that a large majority of the population perceives that the public has little or no influence on policy. This general perception is overwhelmingly unrelated to policy preferences, so it is not a simple matter of disgruntled losers. Similarly, neither participation nor the use of particular modes of participation are major sources of belief in the public's ineffectiveness.

To the extent that the public is perceived as effective, it is because the standard institutions for expressing public preferences—elections and the legislature—are seen as effective agents on behalf of the public. Citizens holding such beliefs are willing to be represented through largely traditional channels and to leave the decisions to traditional agents. They derive no particular sense of efficacy from policy-specific input channels.

In contrast, the vast majority of citizens perceive little current public influence and wish to have more representation through organized public representatives such as groups, petitions, and elections. They also wish to give more decision-making power over policy administration either directly to the public or to direct representatives of the public outside traditional institutional channels (i.e., the general public, citizen advisory committees, and groups). For this group policy-specific citizen participation, although desirable, is not presently perceived as sufficiently influential. Thus, in sum, neither the satisfied nor the dissatisfied currently perceive the policy-specific citizen participation modes as major sources of public influence. The public views the current influence of citizen participation modes as less than it should be and, by and large, is willing to give public representatives broader powers to correct the situation.

Because the primary requisite of a participation mode is that it be effective in translating preferences into policy, we now summarize the findings concerning the perceived effectiveness of the various specific modes. First, none of the modes is strongly perceived as effective, although the legislature, petitions, elections, citizen advisory groups, and, to some degree, interest groups are perceived as, on balance, effective. Petitions, though regarded as effective generally, only carry one-sided preferences, and there is a corresponding division of opinion about their effectiveness: Republicans and individuals supporting recreation see them as effective; Democrats and individuals favoring priority for industrial uses of water see them as ineffective modes for registering public opinion. Although these particular cleavages may be unique to the current Washington water policy arena, the one-sided nature of petitions as a channel for policy preferences suggests that a division of opinion regarding their effectiveness will be a general phenomenon. Similarly, given that legislatures are organized along party lines reflecting some bases of cleavage (and not other cleavages or consensus), the fact that Democrats, pro-agricultural and pro-industrial use individuals, see the legislature as effective and that Republicans and citizens who favor recreation and preservation see the legislature as less of an agent for the general public appears the sort of result that can generally be expected. Interestingly, both the legislature and agencies are seen as noticeably less effective by citizens who have participated in these channels than by nonparticipants. Agencies are further seen as much less effective by pro-recreation individuals, Republicans, conservatives, and voters generally, presumably a partial reflection of the agencies' policy actions.

Citizen advisory committees are notably the least affected by any considerations of policy preferences, party, ideology, or participation. Advisory committees are almost uniformly judged as relatively effective.

With the exception of hearings, which are taken by the public to be more of a forum for voicing specific individual concerns than as an instrument of public representation per se, judgments about the effectiveness of the modes do exert a strong influence on judgments concerning whether they should be used as representation channels. This relationship is most notable in the case of petitions and elections. Apparently, almost any mode that does an effective job will be accepted as a legitimate mode of participation. To a lesser degree in most instances, effectiveness of representation produces support for granting decision-making authority to a channel. Particularly for institutionalized decision-makers (legislature, agencies), effectiveness at public representation would considerably increase support for their decision-making roles. In sum, none of the channels considered here has "the" solution to effective public representation. The drawbacks of each are indicated in the pattern of public judgments. The results generally confirm the public's preferences for (a) a variety of channels, (b) participation that directs the public's specific policy preferences to decision-makers, and (c) any method which "works" in so doing in particular instances.

The recommendations which derive from these results are largely negative, since nothing is working all that well. Citizen advisory groups are the single most useful addition to standard techniques for channelling citizen policy preferences to decision-makers. Probably more important, however, would be revised treatment of citizen communications to the legislature and agencies. To a degree, this appears to involve a deemphasis of more organized transmitters of policy desirability judgments—notably parties and ideology, in favor of the issue-specific judgments of individual citizens. There does not appear to be an institutionalized guarantee, however, that an absence of organized political control would be followed by reaffirmations of direct citizen controls; thus, it would be silly to dismantle existing processes. The real problem is to integrate citizen input into existing processes rather than dismiss the input as often appears to be current practice.

In the water use policy sector the relatively controversial items—such as the recreation, preservation, and industrial use preferences—do not, perhaps paradoxically, appear to be as much of a problem for insuring citizen input as do policies touching on more consensual choices. The controversial items have links to standard modes for organizing policy desirability (parties, ideology), involving participants in standard (parties, legislature, agency) modes of contacting decision-makers, and stimulating participation in politics through either institutionalized or new citizen participation modes. Consensual preferences, on the other hand, tend not to stimulate participation and tend to be underestimated in controversial settings. Thus, transportation, a uniformly low-priority water use, receives more benefits than it should given its overall low priority, while consensually high-priority uses, such as agriculture and domestic use, do not receive representation as high-priority uses.

The problem of identifying and representing consensual policy preferences is not unique to this relatively consensual water policy arena. It is a problem as well in high-conflict arenas such as public education, where consensual preferences (such as placing high priority on teaching basic cognitive skills) may be lost sight of in the heat of political conflict over appropriate educational goals. Designing structures that can detect and keep sight of such consensual preferences in any policy sector is a desirable objective. Once determined, consensual preferences can serve as both a "boundary setter" for political debate and a "priority yardstick" against which to measure other policy proposals. Public choice can then take the form of consensus concerning basic goal X with conflict over goals A, B, and C, rather than, as is often the case, conflict over X, A, B, and C, followed by compromise solutions which sacrifice basic and widely supported objectives.

The registering of consensual priorities and the application of this information to specific policies and programs appears to be one of the foremost potential functions of citizen participation; it would fill a present gap in policy-making without involving many of the previously discussed difficult representation questions applicable to nonconsensual policies. None of the participation modes examined here, with the possible exception of citizen advisory groups, necessarily can carry such consensual, multioption views, however. Perhaps a sequential combination of hearings at the stage of considering policy alternatives, citizen advisory groups during decision-making and basic implementation, then hearings again before implementation—with petitions, elections, groups, decision-maker contacts, and parties as back-up information sources—would best carry consensual preferences through the policy process.

We have found citizen participation in specific policy areas to be a desired and apparently needed channel of representation, though the actual public views in the water policy sector do not closely correspond to the theoretical arguments for and against citizen participation. The practical difficulties, however, are considerable: citizen participation modes are not working much better than institutionalized modes. To the degree that solutions exist, they apply relatively consensual public preferences fairly directly to decision-making, but no single mode of participation does this adequately, so a combination of methods of citizen participation is advisable. The most important considerations are the largest ones and the ones often lost sight of: any participation that translates major public preferences on policy alternatives into actions is desirable; any that does not is unsatisfactory.

NOTES

1. For further description of the study see John C. Pierce and Harvey R. Doerksen (1975).

2. This phrase appears as a standard questionnaire item on survey instruments designed to measure the political distrust of citizens. For example: "Would you say that the government is pretty much run by a few big interests looking out for themselves or that it is run for the benefit of all people?" (Gilmour and Lamb, 1975: 161).

3. These six methods include: signing a petition or initiative, attending a public hearing, contacting or writing a state legislator, contacting or writing a government agency, joining a group, and serving on a citizen advisory committee.

4. Misrepresentation ought to be interpreted with two sorts of sampling errors in mind. Participation itself is a type of sample and so representation is subject to the usual error formulas. Specifically, the greater the number of participants–petitions as compared with citizen advisory committees–the more likely participants are to be representative; the more consensus on an opinion or behavior, the less likely will be overrepresentation in large amounts, while the more infrequent an opinion or behavior, the greater the potential for overrepresentation–recreation use is unlikely to be seriously overrepresented while pro-recreation use priorities are unlikely to be seriously underrepresented; the more variation, the more likely some misrepresentation–rankings of industrial use priorities are more likely to be misrepresented than are transportation use priorities. These sampling error aspects of representation are real problems for politics: misrepresentation is likely when only a few views are heard on issues with a small but intense minority. The use of large-scale participation modes (like elections) to discover a majority on divisive issues is one solution.

The second type of sampling error only concerns analysis, not politics itself. This is the sampling error for our data. Exactly the same formal considerations–sample size, variance, direction of misrepresentation–apply, often in multiplicative fashion, to our errors in perceiving representation and to political errors in representing the public. Therefore, we have tried to be cautious in interpreting our misrepresentation figures, for the same absolute misrepresentation estimate is not equally trustworthy for petitions and for citizen advisory committees. More than single striking scores, we use a pattern of misrepresentation of guide analysis, as sampling error is less likely to produce sensible patterns than to produce a few large "misrepresentations."

REFERENCES

ALESHIRE, R. A. (1972) "Power to the people: an assessment of the community action and model cities experience." Public Administration Review 32 (September): 428-443.

ALTSHULER, A. (1970) Community Control. New York: Pegasus.

ARROW, K. J. (1963) Social Choice and Individual Values. New Haven, Conn.: Yale University Press.

CAHN, E. S. and J. C. CAHN (1968) "Citizen participation," pp. 211-224 in H.B.C. Spiegel (ed.) Citizen Participation in Urban Development. Vol. I. Washington, D.C.: NTL Institute for Applied Behavioral Science.

CLAUSEN, A. R., P. E. CONVERSE, and W. E. MILLER (1965) "Electoral myth and reality: the 1964 election." American Political Science Review 59 (June): 321-332.

DAHL, R. (1967) "The city in future of democracy." American Political Science Review 61 (December): 953-969.

--- (1966) A Preface to Democratic Theory. Chicago: University of Chicago Press.

--- (1962) Who Governs? New Haven, Conn.: Yale University Press.

DAVIS, J. (1973) "Citizen participation in a bureaucratic society: some questions and skeptical notes," pp. 59-72 in G. Frederickson (ed.) Neighborhood Control in the 1970s. New York: Chandler.

DeGRAZIA, A. (1958) "Nature and prospects of political interest groups." Annals of the American Academy of Political and Social Science 319 (September): 113-122.

EDELMAN, M. (1964) The Symbolic Uses of Politics. Urbana: University of Illinois Press.

FANTINI, M., M. GITTELL, and R. MAGAT (1970) Community Control and the Urban School. New York: Praeger.

FRANCIS, W. L. (1967) Legislative Issues in the Fifty States. Chicago: Rand McNally.

FRIEDRICH, C. J. (1941) Constitutional Government and Democracy. Boston: Little, Brown.

GILMOUR, R. S. and R. B. LAMB (1975) Political Alienation in Contemporary America. New York: St. Martin's.

GOLDBLATT, H. (1968) "Arguments for and against citizen participation in urban renewal," pp. 31-42 in H.B.C. Spiegel (ed.) Citizen Participation in Urban Development. Vol. I. Washington, D.C.: NTL Institute.

GRIFFITHS, A. P. (1969) "Representing as a human activity," pp. 133-156 in H. F. Pitkin (ed.) Representation. New York: Atherton.

HERBERT, A. W. (1972) "Management under conditions of decentralization and citizen participation." Public Administration Review 32 (October): 622-637.

LAKOFF, S. [ed.] (1973) Private Government. Glenview, Ill.: Scott, Foresman.

LINDBLOM, C. (1959) "The science of 'muddling through'." Public Administration Review 8 (Spring): 78-88.

LIPSKY, M. (1968) "Protest as a political resource." American Political Science Review 62 (December): 1144-1158.

LOWI, T. (1969) The End of Liberalism. New York: W. W. Norton.

MANN, J. H. and C. H. MANN (1959) "The effect of role-playing experience on role-playing ability." Sociometry 22: 64-74.

MARINI, F. [ed.] (1971) Toward a New Public Administration: The Minnowbrook Perspective. New York: Chandler.

McCONNELL, G. (1967) Private Power and American Democracy. New York: Knopf.

MILLER, S. M. and M. REIN (1969) "Participation, poverty, and administration." Public Administration Review 29 (January-February): 15-25.

PIERCE, J. C. and H. R. DOERKSEN (1975) "Personal values and water resource policy preferences." (unpublished)

PIVEN, F. F. (1966) "Participation of residents in neighborhood community action programs." Social Work 2 (January): 73-81.

POMPER, G. (1975) The Voter's Choice: Varieties of American Electoral Behavior. New York: Dodd, Mead.

ROSE, D. (1976) "Public opinion and water policy," in J. Pierce and H. Doerksen (eds.) Water Politics and Public Involvement. Ann Arbor, Mich.: Ann Arbor Science.

SIGEL, R. S. (1967) "Citizen committees–advice vs. consent." Trans-Action (May): 44-52.

SIMMEL, G. (1950) "The metropolis and mental life," pp. 408-424 in K. Wolff (ed.) The Sociology of Georg Simmel. New York: Free Press.

SMITH, M. P. (1974a) "Elite theory and policy analysis: the politics of education in suburbia." Journal of Politics 36 (November): 1006-1032.

--- (1974b) "The ritual politics of suburban schools," pp. 110-130 in M. Smith et al., Politics in America: Studies in Policy Analysis. New York: Random House.

--- (1971) "Alienation and bureaucracy: the role of participatory administration." Public Administration Review 31 (November-December): 658-664.

SPIEGEL, H.B.C. [ed.] (1968) Citizen Participation in Urban Development. Vol. I. Washington, D.C.: NTL Institute for Applied Behavioral Science.

STOKES, D. E. (1963) "Spatial models of party competition." American Political Science Review 57 (June): 368-377.

STRANGE, J. H. (1972) "Citizen participation in community action and model cities programs." Public Administration Review 32 (October): 655-669.

WILSON, J. Q. (1963) "Planning and politics: citizen participation in urban renewal." Journal of the American Institute of Planners 29 (November): 242-249.

6

Special Districts and Urban Services

ROBERT B. HAWKINS, Jr.

□ CONVENTIONAL WISDOM regarding the optimal structure for the delivery of urban services holds that special districts are not well suited to undertake such tasks.[1] Conventional theories on local government structure view fragmentation as a primary cause of inefficient and unresponsive government at the local level. Consolidation of these units into a unified political and administrative structure is predicted to bring about increased efficiency and responsiveness. From this set of assumptions, special district government is seen as illogical and undesirable, especially in urban areas.

This chapter will outline these contentions and review existing evidence to assess the warrantability of such claims. However, before turning to the theories and evidence, a brief description of special districts is in order. The daily communter across the Golden Gate Bridge interacts with a special district. So does the citizen of San Diego when he or she draws a glass of drinking water from the tap, water that has been transported from the Colorado River by the Metropolitan Water District of Southern California. In the first case the citizen is probably aware of the district because of the highly

AUTHOR'S NOTE: *The research for this study was conducted while I was a Visiting Fellow at the Hoover Institution on War, Revolution and Peace, Stanford, California. The writing took place while I was a Fellow at the Woodrow Wilson International Center for Scholars, Washington, D.C.*

visible conflict surrounding the fare structure and the attempts by the city of San Francisco to dominate the governing board. In the second case most of the conflicts regarding the production and costs of water in southern California have been resolved in the past so the citizen merely presumes that the city of San Diego rather than a district produces the water that he drinks.

These examples demonstrate the use of districts to solve problems requiring large-scale organizations. Districts also are utilized to deal with problems of a different magnitude. Highway City Community Services District was formed in 1969 to provide governmental services to a small community west of Fresno, California, that is primarily Mexican-American with a majority of its 500 families below the poverty line. The district has no full-time staff members so its primary activities are concerned with establishing priorities and with contracting with both public and private agencies for the provision of services. The county of Fresno is the basic provider of most of these services. The district also acts as a forum through which voluntary self-help programs are initiated.

The community of Idyllwild, located in the mountain region of Riverside County, receives its fire and emergency services from the Idyllwild Fire Protection District. In 1971 the district lowered its tax assessment from $1.08 to $0.75 per hundred dollars of assessed valuation. With passage of a state-mandated tax limitation on all local governments, the rate was frozen at the lower figure. Since six of its personnel were funded through the federally sponsored Public Employees Program that terminated in 1972, the district was faced with the necessity of coming up with additional funding to keep its full-time staff operating in order to maintain the desired level of service. A study that would reinstate the old rate of $1.08 was conducted by the fire chief and accepted by the Board of Directors. However, to reinstate this rate it was necessary to place the question on the ballot and to receive a favorable majority vote.

During this same period the Board of Directors considered a proposal from Riverside County which called for the district to contract with the county for provision of fire and emergency services at the lower rate of $0.75. With a turnout of 73.4% of the registered voters, the district's proposal was adopted by a 90% plurality. Two primary reasons were given by the community for choosing the higher rate. First, since a majority of the citizen are retired it was felt that the 24-hour emergency services of the district would provide faster responses than the volunteer services offered by the county.

Second, many felt that the district would be more responsive to their demands than would the county.

Special districts are governments that, like municipalities, derive their primary decision-making capability from state legislative authorization. Districts are used at the local and regional levels by communities of interest to solve public problems of common concern. While most districts have many of the common powers of cities and counties, there are significant differences between the two. Some of these differences are:

(1) Districts frequently perform only one governmental function, such as fire protection, water provision, or street lighting.

(2) Districts can include parts of cities, encompass many cities, or encompass unincorporated areas of counties. The geographic scope of a given function relative to a community of interest is the key factor that determines the boundaries of special districts.

(3) Districts can be governed by city or county elected officials, by representatives appointed by various units of local government, or by independently elected officials.

(4) District revenue patterns differ significantly from those of cities and counties. In California, 54% of the revenues spent by districts are in the form of user fees for services rendered, while cities collected 33% and counties 9% of their revenues from user fees.

(5) Districts in many cases are the direct result of inaction on the part of cities and counties following demands for public goods and services by smaller communities of citizens. Thus, districts, in many cases, represent a very particular interest, whereas cities and counties represent a number of particular interests.

In a basic sense, special districts can best be thought of as governmental entities that are created because of the limitations of general purpose units of local government. This is not either to criticize cities and counties or to suggest that districts are a more rational form of governmental organization. It is, rather, to understand that all organizations have capabilities, as well as limitations. Districts have been organizational solutions used to solve some government problems. From the creation of the Metropolitan Water District of Southern California to providing fire services in remote mountain areas, districts have been used when other delivery systems have been, for a number of reasons, inadequate. Given this use, it is particularly important to understand the capabilities and limitations of districts.

THE TRADITIONAL VIEW OF SPECIAL DISTRICTS

An understanding of special districts has been subsumed under the general theory that has guided not only intellectual inquiry but also reform efforts in the area of local government. This theory, denoted as the reform theory of local government,[2] assumes that because local government is fragmented, it is also inefficient, ineffective, unresponsive, unplanned, and uncoordinated (Committee for Economic Development, 1966: 11-12). A frequently sought remedy is consolidation of this system into a unitary command and administrative organization, predicted to realize the optimal structure for local government. Optimality in this case is defined by such variables as increased efficiency, effectiveness, and citizen satisfaction with local government.

It is from this set of theoretical assumptions that policy makers, in large part, gain their understanding and make recommendations regarding the use of district governments. For example, the Advisory Commission on Intergovernmental Relations (ACIR) finds that "in light of the Commission's *approach to government* (reform theory) . . . it is apparent that many (special districts) have outlived their usefulness. . . ." (ACIR, 1964: 75). The list of particulars that ACIR uses to justify the assertion that special districts have outlived their usefulness are that districts (1) frequently provide uneconomical services, (2) create serious intergovernmental problems, (3) prevent citizens from understanding and controlling them because of their multiplicity, (4) increase the costs of government within an area, and (5) distort the political process by competing for scarce public resources (ACIR, 1964: 74-75).

The Citizens Committee on Local Government Reorganization for the County and City of Sacramento, California, recommended in 1974 that nearly all of the special districts in Sacramento County should be consolidated. They argued (Citizens Committee, 1974: 4-5) that (1) special districts produce disparities in costs of service and a substantial differential in the level of these services; (2) their boundaries have evolved without regard to existing communities of interest; and (3) they impede long-term comprehensive social and physical planning.

Each of the criticisms made by ACIR and the Citizens Committee shares one common theme. Each ascribes limitations in the use of special districts. Such criticisms generally lack empirical verification regarding the capabilities and limitations of special district government. However, such statements do contain propositions that are capable of empirical verification.

While most theorists and practitioners of local government would agree with the preceding arguments, several events in the last decade have caused some to question the validity of such reasoning. For example, if large, integrated organizations are more efficient and responsive than smaller, fragmented units of government, then how does one explain the fact that urban unrest and violence occurred primarily in large centralized cities? Or, how does one explain the fact that minorities made very consistent demands for decentralized forms of neighborhood governments in large cities?[3] In California, it was found that minorities in large cities saw neighborhood government as a more effective way of solving problems with city administrations than did the white majority of those same cities (Haug Associates, 1974: IV–39). Such anomalies have led many critics to rethink the key assumptions of the reform theory.

SPECIAL DISTRICTS AND PUBLIC CHOICE

The public choice approach starts with basic assumptions about individuals, rather than organizations, as the building blocks of political structure. No presumption is made regarding one optimal or rational organizational form for local government structure.

Individuals are assumed to be rational. The usual conditions of rationality are that (1) the individual, because of scarce resources, is forced to make choices; (2) such choices are guided by the individual's desire to maximize those things he values, and (3) the individual can rank his preferences for goods and services (Bish, 1973: 3-13). Individuals are also assumed to have diverse preferences for public goods and services.

Public choice theorists also make assumptions about capabilities and limitations of organizations and about the environments in which they operate. The boundaries of problems vary. Flood control plains, mosquito vectors, and water basins do not respect city or county boundaries. Organizational capabilities are limited by the problems they attempt to solve. A number of public and private water producers can have negative and serious effects on water levels in basins where short-run demands exceed safe supply limits. In such cases voluntary solutions are generally ineffectual; rather, some extraordinary type of organizational response is required. The ability to realize economies of scale also demonstrates variability. The construction of large sewage treatment plants requires greater capital outlays (and different service areas) than do most social services that are labor intensive. Another key constraint is citizen satisfaction with different organizational configurations. Since each of these factors varies, it is expected that if a local government system is to be efficient and responsive there will be a number of governmental entities. Special districts are not seen as illogical units of government but, rather, as organizational choices that may find beneficial use when other institutional alternatives demonstrate operational limitations.

EVIDENCE

Given that the reform theory has been accepted without much question for so long, little evidence about the capabilities and limitations of special districts has been gathered in a systematic manner. A review of the existing evidence will be made in the next section to attempt to assess in a preliminary manner the usefulness of either approach for understanding the capabilities and limitations of special districts.

EFFICIENCY

The traditional mode of analysis in determining the most efficient local government structures has been economies of scale studies. These studies have generally been of two types. First, studies looked at per capita costs in relation to the size of the governmental unit. The second generation of studies has looked at the size of jurisdiction in relation to unit costs for a given service. The notion of economy of scale assumes a decreasing unit cost to increases in scale and, at some point, increasing costs from increased scale.

The general finding of a number of studies using different methodologies has been that there are few, if any, economies of scale in the production of local government services. The only services that are susceptible to economies of scale are those that are capital intensive, such as sewage treatment plants and water transport facilities.[4] In a recent study, ACIR (1970: 2) found that economies of scale were realized by most cities of around 25,000, and that diseconomies of scale began to occur in cities serving populations of over 250,000. This finding generally covers those services that are labor intensive.

Until several years ago there were no studies to compare the relative efficiency of special districts with other units of government. The California Local Government Reform Task Force conducted three studies of the efficiency of districts. First, a comparison of 153 city and special district sewage treatment facilities could find no significant cost differences even though the cost data for cities were systematically understated (Krohm, 1974: 34). A comparison of 100 city and special district fire districts could find no statistically significant relationship between costs even though a weak correlation was found between decreased costs and city provision; again, the cost data for cities were significantly understated (Krohm, 1974). Finally, a study of 144 of California's largest school districts found that there were consistent diseconomies of scale in school systems with student populations of over 2,500 (Niskanen and Levy, 1973: 52-63).

Existing evidence warrants the assertion that special districts are just as capable of realizing efficient operations as are units of general purpose government. It also appears that special districts are just as susceptible to diseconomies of scale as are other units of local government. While economies of scale studies are important, they make certain assumptions that hide two of the most important economic reasons for districts. Returns from scale must, of necessity, assume that there is a suitable environment in which to operate. In many cases, the provision of local government services is in sparsely populated or geographically isolated areas, making provision by a general purpose unit of government highly expensive. In short, there is no way to realize decreasing unit costs.

The question turns to finding the least expensive method of delivery. For example, the state of California studied the feasibility of proving structural fire protection to communities that bordered wildland areas protected by the State Division of Forestry. The primary presumption of the study was that the division's fire fighting system would more efficiently and effectively serve the area than would the numerous districts in various stages of development. What the study found was that it would cost approximately 53% more to provide fire services to these communities by the Division of Forestry. Cost increases were in two areas: (1) increased general costs due to state standards, regulations, and pay scales; and (2) provisions of higher levels of service to some of the communities (Division of Forestry, 1974: 63-64). The study found that different levels of service were a function of "areas in different stages of development and required different levels and sophistication of fire service" (Division of Forestry, 1974: abstract).

The desired level and demand for service demonstrates another important limitation of economies of scale studies. Economies of scale are usually achieved through the standardization of goods and services. By standardizing products, cost variations are eliminated. While such standardizations may decrease the unit cost of goods, they do not occur without other costs to those citizens who desire to consume either more or less of those goods.

For the sake of illustration, let us assume that the fire service to be offered by the Division of Forestry was in the Class 8 range; the best Insurance Services Office (ISO) rating is a one, while the lowest is a 10. Communities in different stages of development can be represented as desiring fire protection in Classes 7, 8, and 9. We will also assume that the level of service provided by the Division of Forestry corresponds to one that would be provided by a group of elected officials attempting to satisfy the median voters (represented by a Class 8 rating) in an effort to gain election. Figure 1 illustrates this relationship.

Line "M" represents the per unit costs and level of services provided by a government unit to the average consumer. The shaded triangles represent the losses incurred by high and low demands for services (Niskanen and Levy, 1973: 1-5). The lower triangle represents the economic losses that a community desiring to consume 9-rated fire service will incur when forced to consume 8-rated services. Likewise, the upper shaded triangle represents the necessary resources that must be contributed by a community to consume a higher level of service than is provided by a 7-rated service. In the first case, citizens pay for more service than they desire, and in the second, more will have to be paid to consume the desired level of service. This situation is not unlike two individuals who respectively want to buy automobiles worth $3,000 and $5,000. If the auto industry only sold cars worth $4,000, each would be worse off than if $3,000 and $5,000 automobiles were also available.

Districts are particularly appropriate institutional mechanisms to reduce these economic losses. By allowing communities with diverse preferences to realize desired levels of service, gains are realized. Whether these districts are adjuncts of cities and counties or run by independently elected boards of directors is a matter of local choice. However, it should be noted that presently no evidence exists

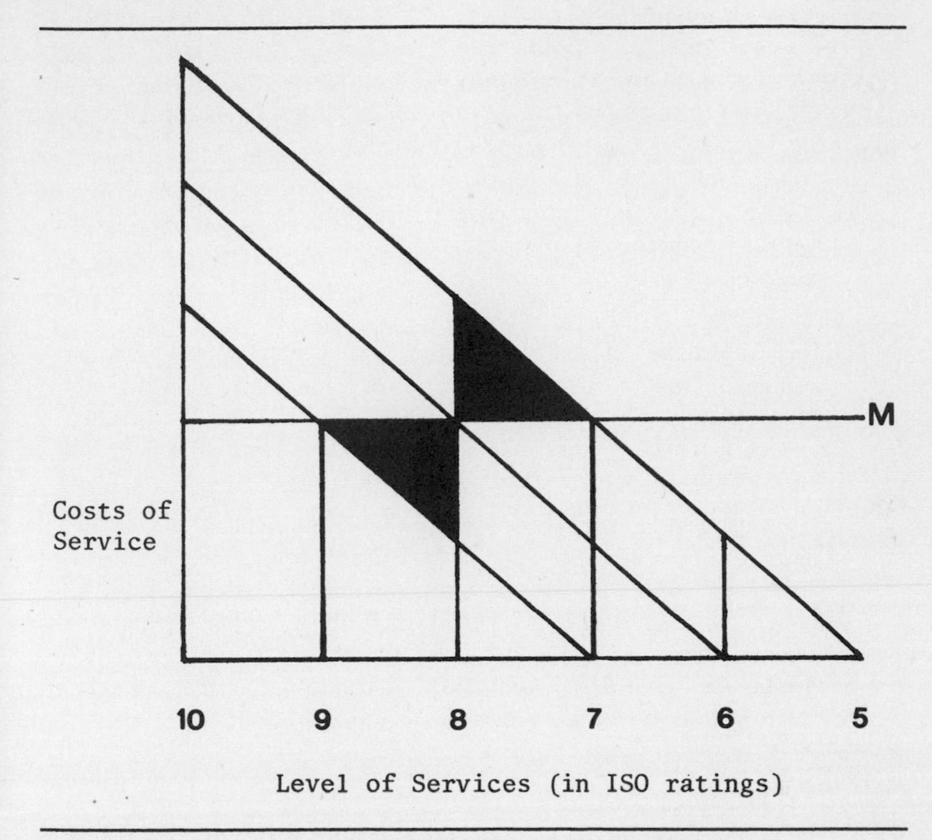

Figure 1: DEMAND OF SEVERAL COMMUNITIES FOR FIRE SERVICE

regarding the comparative efficiency and responsiveness between these two forms of districts.

It is argued that a large number of governmental units in an area will increase the per capita costs. Since districts are usually the largest single type of governmental entity in an area, they draw the lion's share of the blame for these supposed increases. Such reasoning would lead one to predict that as the total number of governmental units for any given metropolitan area increases, there would be a corresponding per capita increase in the costs of local government. Yet, two national studies could find no statistically significant correlation between the number of government units and the total costs of local government. A similar study conducted in California confirms these findings and suggests that costs are actually lower as the number of governmental units increases (Hawkins and Dye, 1970: 17-24; Campbell and Sacks, 1967; Hyman, 1974).

QUALITY OF SERVICES

For many governmental services, quality comes down to the subjective preferences of citizen consumers. Several local government functions have been measured for effectiveness.

(1) A study of city and district sewage treatment facilities was able to control for the quality of output and found no difference between producer efficiency. Comparable quality was achieved at the same average cost (Krohm, 1974: 34).

(2) The comparison of student performance on standardized tests on reading and math at the sixth and twelfth grade levels found that size was negatively related with performance. These findings were associated with school systems serving over 2,500 students (Niskanen and Levy, 1973).

(3) A comparison of Insurance Services Office evaluations of city and district water systems serving fire departments found a weak, positive correlation of .10 in favor of city provision. This finding must be interpreted in light of a bias in favor of city provision. Medium to large size cities, to meet normal water demands, have little trouble meeting ISO water requirements for sustained flows. Smaller cities and water districts usually score lower on such tests.[5]

RESPONSIVENESS

A recent study found that citizens have the following preferences regarding the structure of local government:

(1) Strong desire to maintain voter prerogatives in determining the structure of local government through elections,

(2) Strong preferences for increased decentralization of authority and responsibility to local governing structures, and

(3) Preferences for increased fragmentation rather than consolidation (Haug Associates, 1974: summary).

If such preferences were allowed full expression one would expect more rather than fewer units of local government. This would include more districts. Such an expression would fly in the face of conventional wisdom which holds that local government is unresponsive, in large part, because of its sheer numbers (Committee for Economic Development, 1966: 11-12). Districts are particularly vulnerable to this criticism because they are the most numerous form of local government. Because they are so numerous the argument is made that citizens cannot keep track of them and hence cannot control them. Evidence of this lack of control is inferred from low voter turnout at most special district elections. The logical conclusion of such reasoning is that special districts are unresponsive (Scott and Corzine, 1963; Stark, 1971: 40; Bollens, 1957: 252). While studies have found that districts usually have lower voter turnout than other units of local government, one must not be hasty in drawing the inference that they are necessarily unresponsive (Scott and Corzine, 1963), for voting is only one of many mechanisms that can be used to influence governmental policy.

The key evaluative term is, of course, responsiveness. Responsiveness can and does stand for a number of different attributes of a governmental system. Systems that are inefficient and ineffective in solving problems can be denoted as being unresponsive. In this section our concern with responsiveness centers on questions of citizen satisfaction and citizen perceptions of governmental organization.

Proponents of local government reform have equated responsiveness (as citizen satisfaction) with organizational consolidation. Two aspects of this equation are important. First, it is assumed that because of the multiplicity of governmental jurisdictions, citizens are confused and do not have the resources to make such a system responsive. Consolidation reduces this confusion by creating one administrative organization controlled by one body of elected officials. Second, there is an implicit notion about size in such theorizing. If one organizational structure is to operate effectively and respond to citizen preferences, it must be capable of internalizing all of the significant actions that affect a given area. Thus, for consolidation to be effective it must necessarily include all significant portions of a metropolitan area. Thus, the size factor can range from small semi-rural communities to metropolitan areas such as New York. Since, theoretically, larger integrated cities are more capable than smaller ones of controlling events that affect them, one would expect citizens to perceive larger integrated organizations as nonresponsive.

The key empirical question centers on whether citizens perceive a fragmented system as more confusing than a more centralized, consolidated one. The former theory would predict that one would find less confusion as systems become more centralized. The logic behind this expectation is that a fragmented system of local government is so disorganized that the citizen cannot effectively control and make it responsive. Thus, one would expect citizens to view fragmented government as disorganized and unresponsive and prefer solutions calling for consolidation. Since consolidation efforts have been voted down at a three-to-one ratio during the last 25 years, it is of value to understand citizens' views on local government.

A nationwide study conducted in 1973 by the Harris polling organization for the U.S. Senate Government Operations Committee asked over 16,000 citizens their view on government. One question is most interesting, not only because of the result, but because of its wording. Voters were asked if local government was too disorganized to be effective—a question loaded in favor of the presumption that fragmentation of local government produces undesirable results in the area of effectiveness. However, despite this lack of neutrality, the results shown in Table 1 were obtained (McDavid, 1974: 29).

These findings indicate that residents of smaller communities perceive local government to be less disorganized, a finding particularly interesting when the distribution of districts is analyzed in California. Only 20% of the independent districts are to be found in the 12 most populous counties. A survey of these districts found that over 60% served areas that are less than 40% urban. Similarly, it was found that four out of 10 dependent districts, those controlled and operated by cities or counties, were to be found in urban areas. Thus in California 68%, or 2,875 districts, are to be found in 46 or its 58 counties representing 28% of its population. Apart from some of the large utility and metropolitan districts, the distribution of districts is most likely to occur in suburban and rural areas of the state. This distribution increases the likelihood that citizens residing in small cities and unincorporated urban areas will be served by more units of local government than citizens living in large cities. Reform theory would lead one to expect that citizens served by more units of local

TABLE 1
CITIZENS RESPONSE TO THE QUESTION: LOCAL GOVERNMENT IS TOO DISORGANIZED TO BE EFFECTIVE (in percentages; n = 16,145)

	Total of Sample	Cities 50,000 and over	Suburbs 49,000 or less	Towns (unincorporated) Urban	Rural
Agree	35	41	33	32	32
Disagree	54	49	54	60	58
Not sure	11	10	13	8	10

SOURCE: United States Senate (1973).

TABLE 2
CITIZEN SATISFACTION RELATED TO SIZE OF JURISDICTION
(in percentages; n = 1,507)

	Public Schools	Maintenance of Streets and Roads	Zoning and Land Use	Parks and Recreation
500,000 and over (n = 349)	49	77	46	71
150,000-499,999 (n = 200)	58	76	49	78
50,000-149,999 (n = 301)	66	84	55	79
Under 50,000 (n = 300)	64	84	51	78
Unincorporated (n = 357)	62	77	48	71

government would find it more confusing, an expectation not confirmed by this data (California Local Government Reform Task Force, 1974; Hawkins, 1976).

A survey of 1,500 California voters came up with the following results:

(1) Citizens in smaller communities generally had higher levels of satisfaction with local government services than those in larger communities. This pattern held for schools, street maintenance, police protection, zoning, and parks and recreation (Haug Associates, 1974: IV–4; see Table 2).

(2) Citizens living in small communities were more likely to use their local governments to register complaints about problems.

(3) Minority citizens showed higher levels of dissatisfaction than did the majority respondents in larger cities. Such dissatisfaction was also registered in the higher preferences of minorities for some form of neighborhood government to control the quality of local administrative services (Haug Associates, 1974: IV–43; see Table 3).

TABLE 3
POSSIBLE COURSES OF ACTION TO BE TAKEN IN MAKING LOCAL GOVERNMENT RESPONSIVE TO CITIZEN PREFERENCES, BY ETHNIC GROUP (in percentages; n = 1,507)

	Work With Existing Framework	Form Neighborhood Government	Move Away	Vote Officials Out of Office	Don't Care
Black (n = 116)	37	22	2	35	4
Mexican-American (n = 159)	45	29	2	17	7
Other (white) (n = 1,213)	54	18	1	25	2

Given this evidence, the question turns to an attempt to explain why special districts can be responsive forms of government. In the short space of this chapter, only a few reasons can be offered that require further study. They are:

(1) Districts, more frequently than general purpose units of government, represent communities of highly homogeneous values. Elected officials, therefore, are faced with the relatively easy task of producing goods or services that satisfy the overwhelming majority of citizens.

(2) Districts, because of their generally small size, are relatively easier for citizens to influence when they desire to do so.

(3) Elected officials, because they are required to make decisions regarding only one or two functions, have lower control costs over administrative officials than their counterparts in large centralized political-administrative organizations.

(4) Districts also offer citizens alternative institutional options in providing public goods and services. Such options when not constrained provide a competitive rival to existing local governments which may increase their responsiveness to citizen preferences.

CAPABILITIES AND LIMITATIONS OF DISTRICTS

From existing evidence and experience, one can outline a list of organizational attributes that begin to systematize the limitations and capabilities of special districts.

1. Districts are as capable as other units of local government in realizing efficient operations. Districts are particularly appropriate institutions for providing services where a certain clientele prefers either less or more of public services or goods than are offered by other local jurisdictions. They have also been used effectively in those areas where provision would be more costly by general purpose units of government. Evidence also suggests that districts do not increase the costs of government in an area, or, through competition, distort the revenue distribution process.

2. Districts are just as capable of producing goods and services effectively as are other units of local government. This is particularly true when the scope of a problem extends beyond the boundaries of other units of local government. Districts also allow communities to undertake collective enterprises that require capital investments beyond the capabilities of any one jurisdiction.

3. Districts can also provide highly responsive, as well as unresponsive, governmental services. District government will demonstrate unresponsive governance when statutes controlling their formation and operation fail to guarantee that significant communities of interest are involved, that decision-making rules and regulations do not reflect the interests of these communities, and/or that state authorizations do not allow these communities adequate leeway to determine governing structures that meet their particular circumstances. The diverse set of state authorizations for district government in California is one manifestation of communities of interest desiring different types of organizational structures to meet their particular interests and circumstantial requirements. One of the most pronounced capabilities of districts is their ability to produce different levels and qualities of services in response to community interest.

More work is required to further specify the institutional capabilities and limitation of special districts. It is safe to say that special districts have many positive capabilities that critics have failed to state or understand.

ROLES DISTRICTS CAN PLAY IN THE PROVISION OF URBAN SERVICES

Existing evidence indicates that there is a very important role for districts to play in the delivery of urban services. The extent of this role depends, in large part, upon the principles political architects use in designing such systems.

If we start our design from the assumptions that the key building block of any delivery system is with the citizen as a consumer of public goods and services and that optimality is defined in terms of citizen satisfaction, then a whole host of new options becomes available. Ostrom (1973: 132) has illustrated the difference between this perspective and the reform approach when he states that "a democratic theory of administration will not be preoccupied with simplicity, neatness, and symmetry but with diversity, variety, and responsiveness to the preferences of constituents."

Districts are public instrumentalities that can respond to the requirements of diversity, variety, and responsiveness. The following situations are examples of districts that are particularly appropriate institutional responses to the delivery of urban services:

(1) Where the physical boundaries of a problem extend beyond the boundaries of general purpose local government;

(2) Where significant communities of interest and interdependencies exist and do not conform to existing general purpose boundaries;

(3) Where citizens with strong preferences for services cannot receive them from general purpose units of government;

(4) Where local units of government are unwilling to provide a service because of political reasons;

(5) Where new and complex local government functions require decision makers to understand large amounts of information in order to increase the probability of good results;

(6) Where neighborhoods in large cities cannot receive the level or type of service they desire from large bureaucracies responsible to different constituents;

(7) Where it can be demonstrated that district government can realize economies of scale, more efficient operation, or can capture the necessary fiscal resources to undertake programs;

(8) Where a combination of local, regional, and state interests is necessary to produce an effective public service; or

(9) Where common resource problems require some institutional mechanisms to regulate the behavior of a number of public agencies, both city and district.

SUMMARY

Our information and knowledge about special districts is limited. While a wealth of practical information exists where districts have been used, academic interest has been minimal. Because of this lack of information, an important public institution has not been adequately understood. Existing evidence indicates that districts have an important role to play in the provision of public goods and services. The public choice approach appears to be a particularly attractive theoretical approach to use in expanding our understanding of districts and knowing when their use is an appropriate institutional response to different public problems. The affinity between the demands of minorities for neighborhood government and special districts has never been drawn and is a promising avenue that should be explored.

NOTES

1. The Advisory Commission on Intergovernmental Relations [ACIR] (1964) represents a good cross-section of the type of analyses and recommendations that have prevailed regarding special districts.

2. Warren (1966); see chapters 1 and 2 for an excellent description of the basic tenets of the reform theory of local government.

3. Altshuler (1972) provides a good account of the many facets of this movement.

4. See Alesch and Dougharty (1971) for a good description of the economies of scale concept as it applies to local government; also, see Niskanen and Levy (1973) for a good summary of existing evidence.

5. A Pearson product-moment correlation was used to obtain this finding. The evidence was obtained from ISO ratings of 89 fire departments.

REFERENCES

Advisory Commission on Intergovernmental Relations [ACIR] (1970) "Size can make a difference, a closer look." Bulletin No. 70-8. Washington, D.C.: U.S. Government Printing Office.

--- (1964) "The problem of special districts in American government." Washington, D.C.: U.S. Government Printing Office.

ALESCH, D. I. and L. A. DOUGHARTY (1971) "Economies of scale analysis in state and local government." Report R-748-CIR. Santa Monica, Calif.: RAND Corporation.

ALTSHULER, A. A. (1972) Community Control. New York: Pegasus.

BISH, R. L. (1973) The Public Economy of Metropolitan Areas. Chicago: Markham Series in Public Policy Analysis.

BOLLENS, J. C. (1957) Special District Government in the United States. Berkeley: University of California Press.

California Local Government Reform Task Force (1974) "Public benefits from public choice." (See Sub Report on Special Districts.) Sacramento: Office of Planning and Research.

CAMPBELL, A. K. and S. SACKS (1967) Metropolitan America: Fiscal Patterns and Governmental Systems. New York: Free Press.

Citizens Committee on Local Governmental Reorganization (1974) "Govern together." Sacramento, California.

Committee for Economic Development (1966) "Modernizing local government." New York.

Division of Forestry (1974) "Alternative systems of providing fire protection to life and property on California's privately owned wildlands." Sacramento, Calif.: State Printing Office.

Haug Associates, Inc. (1974) "The attitude of citizens of California towards local government: a primary research project for the California local government reform task force." Los Angeles.

HAWKINS, B. W. and T. T. DYE (1970) "Metropolitan 'fragmentation': a research note." Midwest Review of Public Administration 4, 1.

HAWKINS, R. B., Jr. (1976) "District government in the American political system: the California experience." Draft manuscript submitted for publication. (Because limited copies were printed by the State of California, The Institute for Contemporary Studies, Suite 811, 260 California Street, San Francisco, California 94111 will in March of 1976 have copies of the report available for purchase.)

HYMAN, A. (1974) "Local government boundaries." Sacramento, Calif.: Office of Planning and Research.

KROHM, G. (1974) "The production efficiency of single purpose vs. general purpose government." Findings of the Organizational Structure of Local Government and Cost Effectiveness. Sacramento: California Local Government Reform Task Force.

McDAVID, J. C. (1974) The Major Case Squad of the Greater St. Louis Metropolitan Area. Bloomington, Ind.: Workshop for Political Theory and Policy Analysis.

NISKANEN, W. and M. LEVY (1973) "Suggested changes in the California code to promote responsive and efficient local government." Berkeley: Graduate School of Public Policy.

OSTROM, V. (1973) The Intellectual Crisis of American Public Administration. Tuscaloosa: University of Alabama Press.

SCOTT, S. and J. CORZINE (1963) "Special districts in the San Francisco Bay area: some problems and issues." Berkeley: University of California, Institute of Governmental Issues.

STARK, D. (1971) "Patterns of legislator incumbency in independent taxing non-school districts in California." Ph.D. dissertation. University of Southern California.

United States Senate (1973) Confidence and Concern: Citizens View Americans and Government. Washington, D.C.: U.S. Government Printing Office.

WARREN, R. (1966) "Government in metropolitan regions." University of California Davis: Institute of Governmental Studies.

7

Ballot Reform and the Urban Voter

DONALD G. ZAUDERER

INTRODUCTION

□ IN A DEMOCRACY, governments are supposed to provide goods and services which respond to the needs and preferences of citizens. The task is often complicated by the heterogeneous nature of the population served by political units. Local officials often face a perplexing array of demands stemming from diverse preferences in the community. These demands are processed in the political system and ultimately are translated into public policy. Urban service delivery is the end product of a complex political and administrative process.

Decision-making processes are composed of rules which govern the way political actors compete to influence public policy. The "rules of the game" can give certain actors a competitive advantage in pressing for favorable outcomes. To the extent that rules are unfavorable, individuals may face the prospect of experiencing negative consequences from decision-making in local government.

AUTHOR'S NOTE: *I would like to acknowledge the helpful assistance of Professor Elinor Ostrom at the dissertation stage of this research. Three of my colleagues at the American University, Professors Laura Irwin, David Koehler, and A. Lee Fritschler, made very useful comments on an earlier draft of this paper. I gratefully acknowledge this help, but assume full responsibility for the contents presented.*

There are rules governing public hearings, debate, committee structure and leadership, the use and availability of information, and so on. In addition, there are rules which set the conditions under which individuals compete for elected office. According to Douglas Rae (1971), "Electoral laws are important . . . because they help decide who writes other laws." They also determine the degree to which elected officials are accountable to different interests in the political community.

One electoral mechanism that has received little attention is ballot form. Voting is a process that entails time and effort costs (Downs, 1957; Riker and Ordeshook, 1973). When the costs of voting increase—due to a change in ballot form—certain individuals will have insufficient motivation to participate in the process or to vote as many offices.[1] The structure of the ballot, then, can have an impact on election outcome by influencing voting patterns among different socioeconomic groups. By doing so, it simultaneously affects the representative structure of government. Elections are contests to determine what values predominate in public decision-making. These values are the ones that are translated into public policy—the force which determines the character of urban services.

This chapter examines the consequences resulting from an alteration in the form of the Ohio ballot. In 1949, the voters of Ohio passed a constitutional amendment to replace the single-choice party column ballot with the Massachusetts office-type ballot; the revised ballot contained no provision for a single-choice straight ticket vote. Party emblems (rooster and eagle) headed the columns under the Indiana ballot. Emblems signifying party affiliation were not permitted on the office-bloc ballot. The two ballot forms, then, constitute the independent variables. The dependent variable is vote roll-off, defined as decrease in combined party vote for all offices on the ballot. The consequences of roll-off are analyzed in terms of the possible effects on election outcome and on the representation of minority interests in policy formation.

The study includes election data from 1934 to 1964, and the research structure is an interrupted time series design. It is comparative and includes the analysis of returns from Ohio (all offices with a statewide electorate), the cities of Cincinnati, Columbus, and Cleveland, and the counties of Hamilton, Franklin, and Cuyahoga. In addition, selected wards and precincts in Cincinnati are categorized on the basis of education and race in order to analyze the impact of ballot reform on diverse publics.

RESEARCH ON THE EFFECTS OF BALLOT FORM

There is a small body of research which examines the effect of ballot reform on vote roll-off. A debate in the profession has centered on the extent to which institutional or noninstitutional factors (i.e., party identification) are responsible for higher levels of roll-off. Thus far, scholars have limited the analysis of roll-off to an examination of voting in the upper third of the ballot. In contrast, results presented in this chapter are based on data for all 21 offices (national, state, county) on the Ohio ballot. By analyzing roll-off for all offices on the ballot, one can acquire a more complete understanding of the extent to which an institutional variable such as ballot form is related to the propensity of individuals to vote a complete ballot.

Jack Walker (1966), in an important pioneering effort, examined the impact of the party and office-bloc ballots on roll-off. His analysis, however, did not account for the potential effects of the single-choice straight ticket option, i.e., there was no reference made as to whether or not one could vote a straight ticket with a single "x." In the case of Kansas, he examined roll-off before and after 1912, the date that the office-bloc ballot was introduced. The measure of roll-off was the percentage decrease in total vote from the office of President to the office of U.S. Representative. Walker found that roll-off increased 10% after 1912.

In the case of Ohio, he measured roll-off from the office of President to the office of Secretary of State, and Governor to Secretary of State. An increase in roll-off of two percentage points was noted for presidential years with no significant increase during the off-year general elections. Seventeen office-bloc states were then compared with 17 party column states between the years 1950 and 1962. The comparison was made by studying the decrease in vote between the highest office on the ballot (President or Governor) and the office of U.S. Representative. The mean roll-off figures were consistently higher in party column states, and differential percentages varied from a low of 0.2% to a high of 2.0%.

Walter Burnham has also done considerable research on roll-off, but his analyses are based exclusively on statewide offices. In presenting roll-off figures for Indiana between 1880 and 1968, he found that the lowest percentage roll-off was 0.1% in 1880, and the highest was 6.8% in 1968. The lowest roll-off percentage for Massachusetts was 0.1% in 1887, and the highest was 14.7% in

1900.[2] His analysis, however, does not consider the possible effects of institutional variables such as ballot form or the magnitude of roll-off. The purpose of this chapter is to determine the effects of ballot form on roll-off by analyzing combined party voting for every office on the Ohio ballot over a 30-year time horizon.

SOME METHODOLOGICAL CONSIDERATIONS

Voting returns from selected wards and precincts in Cincinnati were collected, together with data on their racial and educational characteristics. Since wards and clusters of census tracts had boundaries roughly coterminous, it was possible to establish the racial and educational characteristics of individuals living in these political subunits. Still, however, there were a number of judgments that had to be made. What, for example, is sufficient to categorize a ward as being black? The 100% rule would effectively eliminate all possibilities. While almost any choice is somewhat arbitrary, I chose to utilize 90% as the decision-rule.

An additional question relates to the assumption of ward racial character during those election periods that fall between census years. Since there is no means of determining the exact racial composition of wards during the interim years, one is faced with the problem of approximating the racial composition with a minimum of distortion. In dealing with this problem, I assumed a stable racial composition for those elections which *cluster around* the census years, i.e., the four years preceding and subsequent to the census years of 1940, 1950, and 1960. In the case of 1940, the four-year assumption was extended to include the election of 1934. The general practice of utilizing the census year as the midpoint in an eight-year time span decreases the possibility of false categorization —especially given the 90% criterion for determining racial composition.

All wards in Cincinnati were ranked on the basis of mean educational achievement level. As a general rule, wards were placed in the three educational categories (high, middle, low) by simply using the top five, middle five, and bottom five cases from the ward ranking. Wards which were neither 90% black nor white were deleted from the sample. The total ward sampling for the 1940 census was 14, but this was reduced to 12 in 1950 and to 11 in 1960.

After executing the procedures, it became evident that there was no black ward which possessed a mean educational achievement level sufficient to place it in the middle or upper category. This meant, in effect, that it was necessary to move down to the precinct level in order to find a black population which met this criterion. Using census tract data and interviews with precinct captains, it was determined that no black precinct with a high educational achievement level existed until 1960. In terms of the middle category, two precincts were identified for the years 1934-1944, three for the years 1946-1954, and three for the years 1956-1964. These precincts and wards, then, constituted the sample for the study.

BALLOT FORM AND VOTE ROLL-OFF

MEASURING ROLL-OFF

There are two conventional ways of measuring roll-off. One can simply calculate roll-off as percentage of presidential or gubernatorial vote—whichever contest is heading the list of offices. The alternative approach, adopted here, is to utilize the office which received the maximum total vote as the upper standard for calculating vote roll-off. Votes for third-party and independent candidates are included in this calculation. Thus, the vote for secondary offices is calculated as percentage of that office which received the highest total.

The data are organized into four periods, representing equal time spans before and after the alteration in ballot form. The roll-off percentages are then accumulated for equal time periods (before and after ballot reform), and the disparity is displayed in the "differential" columns. Table 1 includes the relevant years which are utilized over the time horizon of this study.

TABLE 1
PRESIDENTIAL ELECTION YEARS

Indiana Ballot	Massachusetts Ballot
Period 1	
1948	1952
Period 2	
1944-1948	1952-1956
Period 3	
1940-1944-1948	1952-1956-1960
Period 4	
1936-1940-1944-1948	1952-1956-1960-1964

TABLE 2
MEAN COMBINED PARTY OFFICE ROLL-OFF FOR PRESIDENTIAL YEARS

	Period 1			Period 2			Period 3			Period 4		
	PRE	POST	DIFF	PRE	POST	DIFF	PRE	POST	DIFF	PRE	POST	DIFF
Ohio	6.6	9.0	2.4	6.7	8.9	2.2						
Cleveland	12.3	19.5	7.2	13.3	19.5	6.2	13.4	22.4	8.0	13.5	21.4	7.9
Cuyahoga County	10.4	15.2	4.8	11.6	15.1	3.5	11.9	16.4	4.5	12.0	16.2	4.2
Columbus	8.3	8.9	0.6	6.9	11.8	4.9	6.8	13.0	7.2	6.5	14.1	7.6
Franklin County	8.1	8.1	0.0	6.6	11.3	3.7	6.6	12.5	5.9	6.3	12.5	6.2
Cincinnati	8.9	14.7	5.8	8.5	13.2	4.7	7.7	13.5	5.8	7.3	14.5	7.1
Hamilton County	8.6	13.1	4.5	8.2	11.9	3.6	7.5	11.0	4.5	7.2	12.4	5.2
High White	7.2	9.1	1.9	6.9	7.5	0.6	6.7	7.8	1.2	6.5	8.7	2.1
High Black							x	14.2	x	x	18.5	x
Middle White	8.6	9.7	1.2	7.6	9.0	1.4	6.9	10.1	3.2	6.6	11.8	5.2
Middle Black	11.8	22.8	11.0	12.7	18.5	5.8	11.0	19.6	8.6	9.7	21.9	12.2
Low White	10.9	21.1	10.1	9.1	19.7	9.8	9.0	20.7	11.7	8.5	22.1	13.6
Low Black	15.4	48.3	32.9	15.2	40.9	25.8	12.4	37.1	24.6	11.4	37.9	26.6

The combined set would represent cumulative roll-off figures in the short run and the long run. The data are interpreted on face value. Both an *immediate* and *consistent* increase in roll-off (subsequent to the alteration in ballot form) would constitute important evidence in support of the hypotheses.

It would indeed be cumbersome to describe the roll-off for all 21 offices. I have chosen to use *mean* roll-off compiled for all statewide offices in the Ohio case, and all 21 offices in the city and county cases. Elections for eight of the nine county offices transpire in presidential years.

MEAN OFFICE ROLL-OFF

The mean value is derived by averaging individual roll-off figures for every office on the ballot. At the statewide level, figures for eight offices are averaged; in the city and county cases, figures for up to 21 offices are averaged.

Table 2 represents roll-off figures for all presidential years since 1936. In the case of Ohio, figures are displayed for just two periods because only a few statewide offices (two in 1960 and three in 1964) were run in the election years which correspond to the last two periods of the time horizon. As of 1960, most statewide offices were extended to four-year terms and were run in nonpresidential years.

In the case of Ohio, the initial differential figure of 2.4% is diminished slightly to the 2.2% level by period 2. This percentage decrease in vote for statewide office amounts to a loss of approximately 78,448 votes. The long-term figures for the cities and surrounding counties are uniformly higher, influenced undoubtedly by the larger number of contests on the ballot. In the case of Cleveland, roll-off increased immediately by 7.2% and the increase has held over time. The increase for Cuyahoga County was 4.8%, and the cumulative figures suggest some consistency as well. Columbus and Franklin County, however, show very little increase in the short run but a rather consistent increase for the ensuing three periods. Cincinnati and Hamilton County experienced an immediate increase, which was sustained over time. Viewing the period 4 figures for the six cities and counties, the increase in roll-off ranged from 4.2% in Cuyahoga County to 7.9% in Cleveland. These figures reflect a very substantial decline in voting for secondary offices.

The data from wards and precincts in Cincinnati reveal considerable variation between educational and racial groups. Dealing first

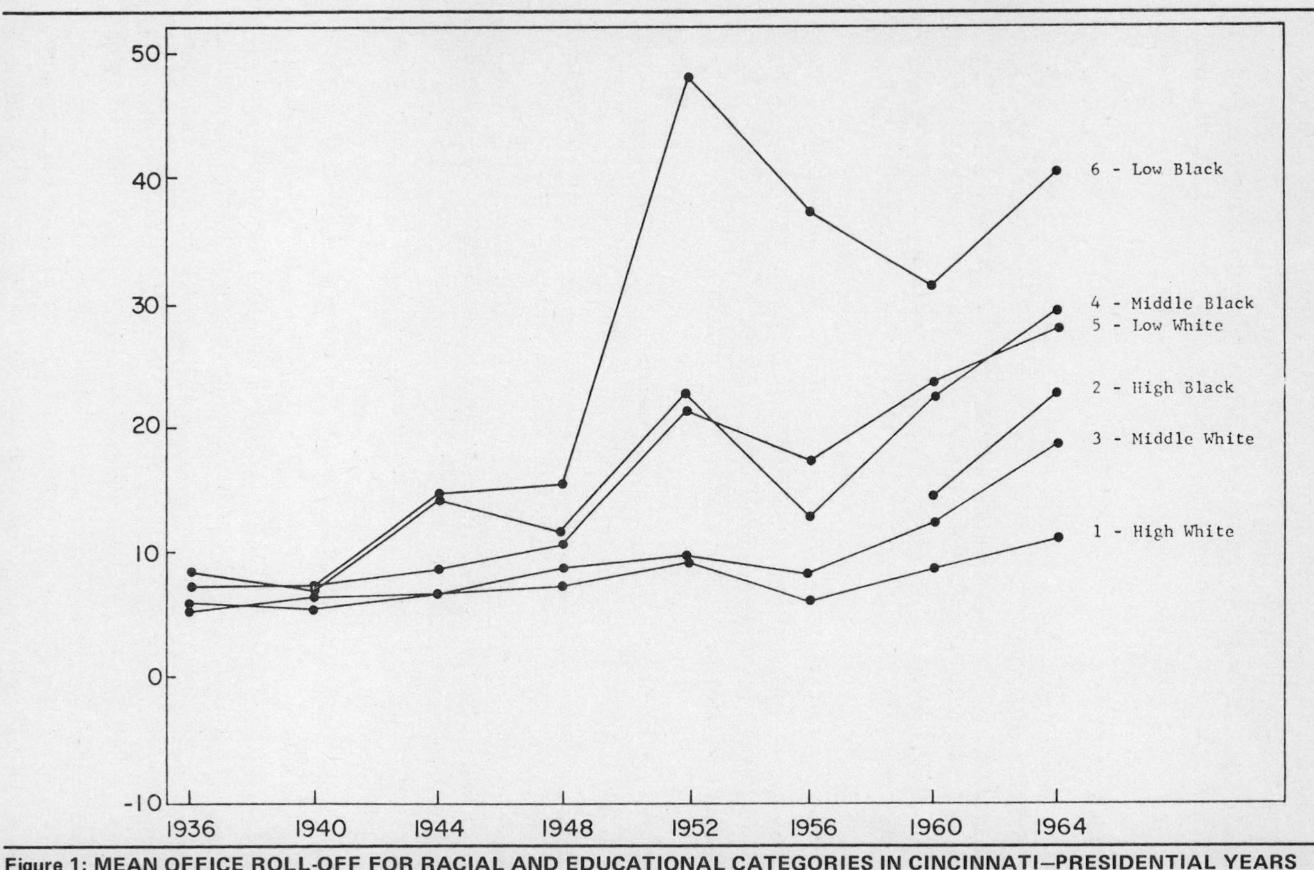

Figure 1: MEAN OFFICE ROLL-OFF FOR RACIAL AND EDUCATIONAL CATEGORIES IN CINCINNATI—PRESIDENTIAL YEARS
(First Presidential Election under the Massachusetts ballot was 1952)

with differential figures, it becomes clear that the initial increase in roll-off for all categories was sustained over time. There is also a progressive increase in roll-off as one moves from higher to lower educational categories. In addition, one finds that within educational groupings (high, middle, low), the magnitude of roll-off is approximately twice as high in the black cases as in the white. For example, a 13.6% figure is registered in the low education white case (period 4), and a 26.6 figure is registered in the low education black case. The roll-off increase in period 4 ranges from 2.1% (high white) to 26.6% (low black). It would appear, then, that the distribution of effects in terms of vote loss falls much harder on the lower status population—both black and white. One might conjecture that higher vote roll-off at the state level is caused by the number of low-status individuals who neglect to vote secondary state offices.

A review of the cumulative (pre and post) figures also generates some interesting relationship results. As expected, the direction of the relationship parallels those which are represented in the differential columns. However, in a number of instances, the roll-off figure in the high education black category is of a greater magnitude than that found in the low education categories. In period 1 (both pre and post) for example, middle education black roll-off is higher than low education white roll-off. Other examples are found in periods 3 and 4, in which roll-off in the high education black category is greater than that registered in the middle education white category. The existence of this phenomenon strongly suggests that race, as well as education, is independently related to roll-off.

An additional consequence of ballot reform can best be demonstrated in graphic form, i.e., as a general trend, the *roll-off spread* between educational and racial categories has increased markedly after ballot reform. Also, the roll-off trends became less stable due to fewer party cues. This is clearly represented in Figure 1, which graphically shows the increase in roll-off after ballot reform—in addition to the marked increase in distance between some categories. The increase in the low education black districts is clearly the most dramatic example of the rise in roll-off due to ballot reform. The presentation of roll-off during presidential years shows that ballot reform effected substantial increases in all jurisdictional, racial, and educational categories. The number of peripheral voters is higher during presidential election years. The excitement and publicity surrounding a presidential contest stimulate higher voting participation. Many individuals who are only marginally interested in politics will still succumb to the stimulus of a presidential campaign. Their general motivation, however, may not be high enough to induce comprehensive voting, i.e., voting a complete slate of candidates. The existence of a significant number of peripheral voters during presidential election years may help explain the rather high roll-off figures.

ROLL-OFF AND VOTE LOSS

An attempt is made here to approximate net vote loss for the state and counties under study. Since it would be very cumbersome to approximate vote loss for every election, I have chosen to utilize the period 4 differential percentage figure (mean roll-off, Table 2) and apply it to the average number of electors who cast ballots in presidential elections from the years 1934 to 1964. This would constitute a summary and representative measure, which would not represent vote loss for any particular office. Table 3 then, represents a rough approximation of vote loss during presidential years.

In the case of Ohio as a whole, the vote loss is approximately 78,448. There were a number of elections where the winning margin for state offices was lower than the vote loss figure for presidential years. The following cases were chosen from a rather cursory look at election data:

1940	Auditor	winning margin: 45,999 votes
1944	Secretary of State	winning margin: 57,135 votes
1948	Treasurer	winning margin: 32,345 votes
1952	Lieutenant Governor	winning margin: 41,753 votes

It is clear that the decrease in participation for secondary state offices is potentially significant, i.e., many elections have been won by fewer than 78,448 votes.

The winning margins for elections in Hamilton County and Franklin County were also examined. In the case of Hamilton County, the vote loss figure was 17,256, and there were at least 18 cases where the margin of victory was substantially lower than this figure. This was also determined by a rather cursory examination of election data. Utilizing the same procedure for Franklin County (vote loss 13,766), 23 cases were found in which the winning margin was lower than the *average vote loss.* Thus, it is quite clear that the magnitude of vote loss is substantial enough that some contests

TABLE 3
INDEX OF VOTE LOSS FOR PRESIDENTIAL YEARS

Presidential Years	$\bar{X}$ Vote Turnout		% Roll-Off		Vote Loss
Ohio[a]	3,565,810	x	.022	=	–78,448
Cuyahoga County	608,554	x	.042	=	–25,559
Franklin County	222,031	x	.062	=	–13,766
Hamilton County	331,849	x	.052	=	–17,256

a. The Ohio roll-off percentage (presidential years) is based upon the period 2 differential figure shown in Table 2. It will be recalled that, as of 1960, very few state offices (one in 1960 and two in 1964) have been presented to voters in presidential years.

might have been reversed had the participation level not decreased. Furthermore, from data on Cincinnati wards and precincts, it is clear that roll-off was greater for the low education groups. It is reasonable to presume that, had these individuals voted for secondary offices, Democrats would have received the lion's share of support. It is clear that, when one analyzes roll-off for *all* offices on the ballot, the decrease in voting is far more substantial than that described in previous research on the subject.

BALLOT FORM AND URBAN SERVICES

Ballot reform in Ohio has led to a substantial decrease in voting for offices such as County Commissioner. The commissioner contest is positioned 13th on the ballot. The powers and prerogatives of this office are significant and range from supervising all other county offices to overseeing public welfare policy and administration. The decline in voting for the office of commissioner among low educated whites was 18%; a decline of 31% was registered among low educated blacks. It is significant to note that individuals most likely to be on some form of public assistance are often neglecting to vote for the office which is responsible for providing welfare services.

Elected officials in the United States must periodically go before the public and be judged on their worthiness to continue in office. Theoretically, this requirement encourages representatives to respond to the needs and preferences of their constituencies. Elections, after all, are instruments of accountability. Ballot reform in Ohio, however, has led to a substantial decrease in voting for secondary offices among low educated citizens. This group, then, has become less vital to the election and reelection efforts of candidates. Under this circumstance, it may become less rational for an office holder to consider the needs and preferences of this public—because the benefits in terms of future support may not be of a sufficient magnitude to warrant the expenditure of time in their behalf. By working in behalf of low educated publics, one foregoes the opportunity to gain the support of wider publics who are more likely to turn out consistently and vote a complete ballot. While office-holders may be motivated by ideology and a sense of mission, they nonetheless need to pay close attention to their own survival. Consequently, future vote support is an important consideration in formulating policy positions. If low educated blacks and whites are less likely to vote country offices, one might speculate that their interests would be less firmly defended in the decision-making process. The power of low socioeconomic groups to influence the delivery of urban services may have been further diminished by ballot reform in Ohio.

NOTES

1. For further elucidation of this argument see Zauderer (1971: 47-58).
2. See Burnham (1970, 1965).

REFERENCES

BURNHAM, W. (1970) Critical Elections and the Mainsprings of American Politics. New York: W. W. Norton.

--- (1965) "The changing shape of the political universe." American Political Science Review 59 (March): 9.

DOWNS, A. (1957) An Economic Theory of Democracy. New York: Harper & Row.

RAE, D. (1971) The Political Consequences of Electoral Laws. New Haven, Conn.: Yale University Press.

RIKER, W. and P. ORDESHOOK (1973) An Introduction to Positive Political Theory. Englewood Cliffs, N.J.: Prentice-Hall.

WALKER, J. (1966) "Ballot forms and voter fatigue." Midwest Journal of Political Science (November).

ZAUDERER, D. G. (1971) "The rig of the game: an analysis of the political consequences resulting from an alteration in the Ohio ballot form." Ph.D. dissertation. Indiana University. (unpublished)

8

Solid Waste Collection in Metropolitan Areas

E. S. SAVAS

□ PRODUCTIVITY IN THE DELIVERY of municipal services is a major concern in local government. Attention has been focused, therefore, on numerous factors which affect the performance of these services. One of the most important issues is the institutional arrangement by which municipal services are delivered.

Many institutional arrangements exist in the United States for the delivery of these services. County and municipal agencies provide services, and so do private firms, which may sell services directly to residents under exclusive franchises or under conditions of open competition. Governments purchase services from private firms and from other governments. A study by the International City Management Association (1964) showed that three-quarters of the 1,007 reporting cities contracted for one or more municipal services. Special districts have been created to satisfy particular service needs. Health services, social services, transportation, and education are provided under a variety of such arrangements, as are water supply,

AUTHOR'S NOTE: *I wish to thank Professor Barbara J. Stevens, Eileen Brettler Berenyi, Daniel Baumol, and Chris Niemczewski–all associated with the Center for Government Studies at Columbia University's Graduate School of Business–Public Technology, Inc., and the International City Management Association for their assistance in carrying out the work reported here. The support of the National Science Foundation, under Grant No. SSH74-02061 A01, is gratefully acknowledged.*

street cleaning and maintenance, installation and maintenance of street lights and traffic lights, refuse collection and disposal, vehicle maintenance, and the vital emergency services: police, fire, and ambulances.

The question arises quite naturally as to the relative efficiency and effectiveness of alternative service arrangements. Downs (1970), Savas (1971), and Niskanen (1971) have called attention to the monopolistic behavior of governmental bureaucracies. "Reprivatization" has been examined and advocated (Drucker, 1968; Savas, 1971; Fitch, 1974), and strategies for introducing competition —public and private—in the delivery of municipal services have been analyzed (Savas, 1974). Nevertheless, despite this attention there has been little in the way of definitive research in this area. There have been many particular studies of individual services in specific localities, of course, and much in the way of polemic tracts and unsubstantiated assertions, but the basic policy question remains unanswered: How does the institutional structure of urban services affect their efficiency and effectiveness?

The Center for Government Studies in Columbia University's Graduate School of Business has been examining these issues through a large-scale study, nationwide in scope, of a particular municipal service: solid waste collection and disposal. The starting point in this study is to create a suitable conceptual framework for classifying alternative institutional structures used to deliver this service and then to measure the extent to which the different structures are employed.

INTRODUCTION TO SOLID WASTE MANAGEMENT

Because of their vital role with respect to public health, solid waste collection and disposal are the concern of government. Generally speaking, they are a responsibility of local government, although states and the federal government are paying increasing attention to these areas and establishing standards through legislative action. While waste collection and disposal do not share the life-saving characteristics of the local emergency services—police, fire, and ambulances—they have even higher political visibility because garbage requires conscious action every day by every family. If service is unsatisfactory, the fact is quickly evident. The importance of this subject to local government is demonstrated by a 1974 survey in which elected local officials named solid waste management most frequently as a problem, ranking it ahead of even crime and housing (National League of Cities, 1974).

But in addition to the active involvement by local government in solid waste management and the relatively high incidence of waste collection by municipal agencies, the private sector, too, plays a significant role in this service. For these reasons, namely a vital service, high visibility, and extensive use of public and private service delivery systems, as well as the *relative* ease of measurement (at least by comparison to education and social services, for instance), solid waste provides an excellent environment in which to study the fundamental policy issues identified here.

DEFINITION OF SOLID WASTE

The conceptual, and widely accepted, definition of solid waste is provided by the American Public Works Association (APWA, 1975). Wastes are defined as "useless, unwanted or discarded materials," and include solids, liquids, and gases. Those wastes which are solid are referred to as solid wastes or refuse. The two terms are used synonymously.

It is symptomatic of the primitive state of analysis of the subject that this commonly accepted definition is quite inadequate. To begin with, uselessness, like beauty, is in the eye of the beholder. The livelihood of more than a million people is dependent upon such useless material. Increasingly, useful material and energy are being recovered from it. Furthermore, much useless material is not discarded but remains in attics, garages, and generally underfoot, that is, not in the solid waste stream. Indeed, according to the APWA definition, a typical suburban garage sale could be described somewhat whimsically as a redistribution of solid waste.

As for being "unwanted," it should be noted that parents are forever discarding their children's most prized and wanted possessions, albeit after contemptuously calling it "junk." In counterpoint, much unwanted material remains undiscarded, simply because the "cost" (usually the effort) of disposal exceeds the "cost" (usually effortless) of retention.

In light of these problems, the following conceptual definition of solid waste is offered: *solid material which is discarded.* This simple definition ignores the irrelevant issue of the usefulness, value, or desirability of the matter in question, but, inasmuch as discarding is

TABLE 1
CLASSIFICATION OF SOLID WASTE

Type	Composition (Descriptive)	Source
1. Agricultural waste		
a. Crop residues	harvesting residue, vineyard and orchard prunings, greenhouse wastes	farms
b. Animal	manure, slaughterhouse wastes	farms, feedlots, slaughterhouses
2. Mineral waste	earth and rock from mining, extraction, and refining	mines, ore processing and mineral refining plants
3. Municipal solid waste		
a. Garbage	waste from the storage, handling, sale, preparation, cooking, and serving of food	households, institutions, and commercial establishments
b. Rubbish (or trash)		
i. Combustible (mostly organic)	paper, cardboard, wood, plastics, rags, cloth, leather, rubber, yard waste (grass, leaves, and the like)	same
ii. Noncombustible (mostly inorganic)	metal, cans, metal foil, dirt, stones, crockery, ceramics, glass, bottles, and the like	same
c. Ashes	residue from fires used for cooking and for space heating	same
d. Bulky waste	stoves, refrigerators, heaters, other large appliances, furniture, crates, tires, auto parts, tree limbs, and the like	same
e. Other municipal waste	street and alley sweepings, catch-basin dirt, contents of litter receptacles in public places, refuse from parks and beaches, dead animals, tree and landscaping refuse (other than yard waste)	streets and other public property
4. Abandoned vehicles	automobiles and trucks	same
5. Industrial waste	waste from industrial processes, manufacturing, and power generation, including cinders, ash, scraps and shavings of wood, metals, plastic, and paper	factories and industrial plants
6. Construction and demolition waste	lumber, concrete, plaster, roofing, pipe, brick, conduit, sheathing, wire, insulation, and the like	construction sites
7. Hazardous waste	pathological waste, explosives, radioactive material, poisons, hazardous chemicals, pesticides	industry and institutions
8. Sewage treatment residues	screenings, grit, digested and dewatered sludge	sewage treatment plants

an intentional act, it implies that the discarder judges the material to be of *relatively* little value to *him.*

Table 1 describes and classifies the various kinds of solid waste and also categorizes them by composition and by the source of generation.

GENERATION AND DISPOSAL OF SOLID WASTE

Of the solid waste generated each year, very little is collected for subsequent disposal. A disproportionate share of the refuse that is not collected consists of agricultural and mining wastes which are deposited in the vicinity of the sites at which they were generated. Other wastes which never enter the collection system are discharged into waterways, incinerated at the site of generation, or discarded as litter which is never collected.

The relatively small fraction of solid waste which is collected consists of municipal waste (including litter), industrial waste, construction and demolition waste, and sewerage sludge.

Of the refuse collected, most is disposed of at open dumps, a practice that is being phased out slowly. The remainder goes to sanitary landfill sites and, to a limited extent, to incinerators. The amount which is recycled is negligibly small, at present, in terms of weight.

AMOUNTS OF SOLID WASTE

Estimates of the amounts of solid waste generated annually, by source, are confusing, inconsistent, and contradictory. Only rough approximations are available, and these are offered to provide a sense of the order of magnitude of solid waste management activities.

It has been estimated that somewhere in the vicinity of four billion tons of solid waste are generated annually in the United States. Of that, about 90% is agricultural and mineral waste, which is not subject to collection. Table 2 presents data on the amount of solid waste, by type, generated in the United States. The classification follows that in Table 1.

Various estimates of residential solid waste generation appear to be reasonably consistent at about 90 million tons annually, or 2.4 pounds per capita per day. The amount of municipal solid waste, which includes residential, commercial, and institutional waste, as well as street cleanings, and the like (see Table 1) is in the range of

TABLE 2
SOLID WASTE GENERATION IN THE UNITED STATES[a]

Type of Solid Waste	Millions of Tons Per Year	Lbs/Person Per Day
Agricultural[b]	2,100 – 2,300	55.6 – 60.9
Mineral[b]	1,100 – 1,700	29.1 – 45.0
Municipal[c]	127 – 186	3.36 – 4.92
Residential	90.3	2.39
	(87 – 102)	(2.3 – 2.7)
Garbage	17.8	0.47
Rubbish (excluding yard waste)	48.0	1.27
Yard waste	18.1	0.48
Bulky waste	6.4	0.17
Commercial and institutional	34.7	0.92
	(34 – 62)	(0.9 – 1.7)
Other municipal waste	6 – 22	0.17 – 0.61
Abandoned vehicles[d]	1.0	.03
Industrial	34 – 105	0.95 – 2.90
Construction and demolition	5 – 19	0.13 – 0.53
Hazardous	NA	NA
Sewage treatment residues[d]	22	0.61

a. Unless otherwise indicated, all estimates are from Smith (1975).
b. Lower estimates are from the Council of State Governments (1973); upper estimates are from Council on Environmental Quality (1970).
c. Data for all components of municipal solid waste are for 1971.
d. Public Health Service (1968).
NA = not available.

127 to 186 million tons annually; the large range is due primarily to uncertainties in the amount of commercial and institutional waste.

SOLID WASTE COLLECTION

The importance of municipal solid waste, in terms of public policy, derives from its presence in urban areas: it is found where people are, and the more people, the more municipal solid waste; furthermore, the greater the population density, the greater the impact of solid waste on public health and environmental quality, the greater the cost of service, and the greater the difficulty of disposal.

Municipal solid waste can be removed from the premises where it is generated in several ways: (1) it can be ground up by garbage-disposal units ("garbage grinders") into a slurry and discharged into the waste-water system; (2) it can be conveyed pneumatically through tubes to a central collection point (this is a recent technological development, particularly applicable to densely situated multiple dwellings); or (3) it can be placed in suitable

containers, which are emptied into or carried away by an appropriate vehicle. (Street sweeping fits into this category, if one considers the hopper of a mechanical sweeper as the container in question.) The first two methods are insignificant in terms of the total amount of solid waste removed; the third method is the only important means of removing municipal solid waste and is the one addressed here.

CLASSIFYING SOLID WASTE COLLECTION SERVICES

SERVICE ELEMENTS

The organizational structure for delivery of solid waste collection services can be specified by four elements or parameters:

(a) the service recipient
(b) the service provider
(c) the service arranger
(d) the service type.

Each of these four is discussed in turn.

(a) Service Recipient. The recipient of solid waste collection service is either (1) a commercial establishment, such as a restaurant, office building, store, or supermarket; (2) an institution, such as a school, hospital, jail, or nursing home; (3) an industrial establishment, such as a steel mill or factory; (4) a residential household, which may be in either a large multiple dwelling or a small structure of one to several housing units—most cities divide their residential structures into these two classes and apply different collection arrangements to them; or (5) the streets and other open places in a community, which can be said to "produce" the component of municipal solid waste identified above as litter and street sweepings.

(b) Service Provider. The service provider is the agent who actually collects solid waste from the service recipient, i.e., removes it from the premises. For municipal solid waste the following six important service providers can be identified:

(1) A local government agency, often within a sanitation department or a department of public works, of the jurisdiction (municipality or township) in which the service recipient is located.

(2) A unit of a general-purpose government other than the municipality or township in which the service recipient is located, such as a county government unit or even a unit of another municipality.

(3) Private firms engaged in the business of collecting solid waste. Such firms are identified in various ways in various places. Most often they are called garbage, refuse, rubbish, trash, or waste haulers or contractors. Sometimes they are called carters or scavengers.

    It is necessary to distinguish between a private firm that has a territorially exclusive franchise and one that does not. The former is referred to as a franchisee or a franchised firm. The distinguishing characteristic of such a firm is that service recipients (or a given class) located within the franchise territory act as service arrangers but are not permitted to select any other organization to provide service delivery. Indeed, if the locality has a mandatory collection law, which requires each waste generator (service recipient) to purchase regular service from a service provider, the service recipient may not even act as his or her own service provider (see below) but must arrange for service from the franchise for this work.

(4) A special district or authority created by state or local law.

(5) The service recipients themselves; for example, in many communities individual householders carry their garbage and trash to the local disposal facility. Similarly, a supermarket chain might assign a truck and several employees to service the units of the chain that are located within a reasonable area.

(6) A voluntary association; for instance, a neighborhood civic association, while not a formal unit of government, might collect that neighborhood's street trash when a more formal arrangement does not exist.

(c) Service Arranger. The concept of service arranger is less obvious than the concepts of service recipient and service provider and requires careful explication. The service arranger is the agent who assigns the service provider to the service recipient. The four important service arrangers for the collection of municipal solid waste are:

(1) The local jurisdiction (municipality or township or county) where the service recipient is located; the jurisdiction may decide to use a

government agency or a private firm (as service provider) to collect its residents' solid waste.

(2) The service recipients themselves, as in the case of residents or commercial establishments that decide to hire a private firm to haul away the trash.

(3) A special district or authority, which, like a municipality, may choose to use its own work force or a private firm as the service provider for the service recipients within its jurisdiction.

(4) A voluntary association may, as above, arrange for its own collecting, or may contract with a private hauling firm to collect solid waste from the homes or businesses of members of the association. The latter is not uncommon in unincorporated areas.

(d) Service Type. The final parameter needed to define the organization pattern is the type of service. This term refers to the type of solid waste collected. Garbage (i.e., putrescible waste), dry trash, yard trash, bulk waste, leaves, newspapers, and mixed residential refuse are examples of types of residential solid waste, and different service providers and arrangers can be used for the different types of service. It would not be surprising, for example, to have the municipality as service arranger and provider for residential garbage and dry trash collection in a community, while the service recipient acts as service arranger and hires a private firm as service provider for the collection of bulk waste and, at the same time, acts as his or her own service provider and brings yard trash to a composting or disposal site. Furthermore, the homeowners may authorize the local Boy Scout troop, a voluntary association, to collect their old newspapers for recycling. For commercial establishments and institutions it may also be useful to distinguish among certain types of service: garbage, corrugated paper, tabulating cards, combustible trash, and so on.

Table 3 summarizes the different elements of this organizational framework. It is clear that a four-dimensional matrix, one dimension for each of the four parameters discussed above, is needed in order to describe completely a community's organizational pattern for collecting municipal solid waste. It is therefore obvious that the situation is considerably more complex than the simplistic, one-dimensional classifications of communities as municipal, contract, or private (and combinations of these) that have been conventionally used heretofore (American Public Works Association, 1966). For example, a city described by the American Public Works Association

TABLE 3
ORGANIZATIONAL ELEMENTS IN SOLID WASTE COLLECTION

Service Recipient	residential household (multiple dwellings and the like)
	commercial establishment
	industrial establishment
	institutions (schools, hospitals, religious structures, and the like)
	streets (roadways, sidewalks, litter baskets)
	other public places
Service Provider	local government agency
	another local government
	private hauling firm
	special district or authority
	service recipient
	voluntary association
Service Arranger	municipality
	special district or authority
	service recipient
	voluntary association
Service Type	mixed residential refuse
	garbage
	trash
	yard trash
	combustibles
	noncombustibles
	bulk
	paper
	cans

NOTE: This is not necessarily an exhaustive list, but it includes the major elements that have hitherto been identified.

as having "municipal and private collection" could be one where the service is provided to all one-family dwellings by a municipal department, while all other collection is carried out for a fee by a single private firm which has an exclusive long-term franchise. On the other hand the same term, "municipal and private," would be applied under the APWA classification scheme to a city where a municipal department collects only residential garbage once a week in one area of the city while several private firms compete vigorously and offer a wide array of services to all other households and to all commercial establishments.

SERVICE ARRANGEMENTS

As shown in Table 3, there are six major kinds of service providers, and four principal service arrangers. In principle, therefore, 24 different pairings could exist. However, as shown in Table 4, five of

TABLE 4
SERVICE ARRANGEMENTS FOR SOLID WASTE COLLECTION

	Service Provider				
Service Arranger	Local Municipality	Other Government	Private Firm	Special District/ Authority	Service Recipient
Local Municipality	municipal (New York)	intergovern-mental contract (Lakewood, Calif.)	contract (Boston)	municipality–special district (Des Moines)	NA
County Special District/ Authority	NA	NA	special district contract (Jersey City)	special district (Indianapolis)	NA
Service Recipient			franchise (San Fran-cisco) or private (Portland, Oregon)		self-service (Eugene, Oregon)
Voluntary Association			association contract (Houston)		NA

NA = not applicable.

the conceivable pairings (hereafter called "arrangements") are essentially meaningless or impractical (marked "not applicable"), and in fact only nine are found to any significant degree. Distinguishing between franchised and nonfranchised firms and between municipal and county governments results in the addition of two more arrangements, for a total of eleven. These eleven are named in Table 4 and listed in Table 5. As is discussed below, the first five arrangements listed in the table—municipal, contract, franchise, private, and self-service—are the common ones and account for the overwhelming majority of the arrangements in effect for residential, commercial, institutional, and industrial solid waste collection. Table 4 also identifies—for illustrative purposes—specific cities which utilize the indicated arrangement, at least to some extent, for the collection of residential refuse from one-family dwellings.

There may be a certain amount of ambiguity as to the identity of the service arranger when the service provider is a private firm, particularly in the case of residential collection, inasmuch as both the municipality and the service recipient (household) may play a role. Three common arrangements—municipal, contract, and franchise—can be distinguished by reference to three conditions: the presence

TABLE 5
NOMENCLATURE AND DEFINITION OF
COMMON SERVICE ARRANGEMENTS

Service Arranger	Service Provider	Name of Arrangement
municipality (or township)	municipality (or township)	municipal
municipality (or township)	private firm	contract
service recipient	private firm with exclusive franchise	franchise
service recipient	private firm without exclusive franchise	private
service recipient	service recipient	self-service
county	county	county
municipality (or township)	another government	intergovernmental contract
special district (or authority)	special district (or authority)	special district
special district (or authority)	private firm	special district—contract
municipality (or township)	special district (or authority)	municipality—special district
voluntary association	voluntary association	voluntary association
voluntary association	private firm	association contract

or absence of a contractual relationship between the municipality and the private firm; whether or not collection service is mandatory; and whether the municipality or the firm collects money from the service recipient (household) for the service. Figure 1 defines the various conditions and the resulting arrangements.

The entries in Table 4 and the third column in Table 5 give the names that will be used throughout the report as a shorthand description of the indicated pair of service arrangers and providers. In other words, reference will be made, for example, to municipal or contract arrangements, or to municipal or contract collections, or to municipal or contract service. In some cases the names given are somewhat ambiguous (for example "private") but they have acquired recognition and acceptance through prior usage. Therefore they have been retained here, with some reluctance, in order to avoid confusion and to facilitate comparison with prior studies. Because of the careful way that service arrangements have been defined here, no confusion should result.

MULTIPLE SERVICE ARRANGEMENTS

It should be noted that multiple arrangements can exist within a community—that is, even if attention is restricted to a single class of service recipient (the household) and a single type of waste (mixed residential refuse), several different arrangements can coexist. In

<table>
<tr><td colspan="2" rowspan="3"></td><td colspan="3">Who collects money from service recipient?</td></tr>
<tr><td rowspan="2">Firm</td><td colspan="2">Municipality</td></tr>
<tr><td>Service is not Mandatory</td><td>Service is Mandatory</td></tr>
<tr><td rowspan="2">Is there a contractual relationship between the municipality and the private firm?</td><td>Yes</td><td>Franchise</td><td>Franchise</td><td>Contract</td></tr>
<tr><td>No</td><td>Private</td><td colspan="2">Not Applicable</td></tr>
</table>

Figure 1: CHART SHOWING DEFINITIONS OF CONTRACT, FRANCHISE, AND PRIVATE ARRANGEMENTS WHERE THE SERVICE PROVIDER IS A PRIVATE FIRM

Indianapolis, for example, four arrangements can be found, as shown by the four entries in Table 6, which is an abbreviated version of Table 4. (Because our attention at this point is focused on only two of the four dimensions, a two-dimensional table suffices.) A special authority, the Indianapolis Sanitary District (ISD), has its own work force and collects waste from most households but also contracts with private firms to provide service to households in certain areas of the city. Nevertheless, there are many households for which ISD is not authorized to act as service arranger; they lie outside the ISD territory but inside the city limits. Those households are free to hire private firms or to carry their own refuse themselves to a disposal facility, and both of these arrangements are found.

Table 6A depicts the various arrangements employed in Indianapolis; Table 6B goes further and portrays the extent of use of each arrangement. This kind of quantitative information is usually difficult to obtain, however, for while many cities have multiple arrangements, there is considerable uncertainty in such cities about the number of service recipients serviced by each arrangement; generally speaking, public officials do not have complete information about this. Therefore, where such estimates are presented below, they should be interpreted as just that—estimates.

It should be emphasized here that the presence of multiple arrangements for the same class of service recipient and the same type of waste, as in Indianapolis, in no way should be interpreted as

TABLE 6
MULTIPLE SERVICE ARRANGEMENTS IN INDIANAPOLIS[a]

A. Incidence of Arrangements

	Service Provider		
Service Arranger	**ISD[b]**	**Private Firms**	**Service Recipient**
ISD[b]	X	X	
Service Recipient		X	X

B. Extent of Arrangements[c]

	Service Provider			
Service Arranger	**ISD[b]**	**Private Firms**	**Service Recipient**	**Totals**
ISD[b]	70	7		77
Service Recipient		18	5	23
Total	70	25	5	100

a. For collection of mixed residential refuse from households.
b. Indianapolis Sanitary District
c. Figures represent the percentage of all households in the city.

poor organization, a confused situation, or a bad condition. In fact, one can offer the reasonable hypothesis that the existence of alternative arrangements in the same city results in healthy competition and provides standards for comparisons that are lacking in cities with only a single arrangement.

Multiple arrangements can also be found with respect to commercial establishments. For example, in some cities a municipal agency collects from stores and offices on city streets but not from similar establishments located in shopping centers; the latter must make other arrangements, either for service by a private hauler or for self-service. A similar situation prevails in a community where the municipality collects from stores, but only twice a week, for example. This may be sufficient for some establishments (e.g., dry cleaning) but not for others (e.g., restaurants). The latter would then make other arrangements for the higher frequency of service that they require.

SERVICE PATTERNS

At this point, having defined the elements of service structures, and having defined an arrangement as a service arranger-provider pair, and having noted the use of multiple arrangements, it is possible to synthesize a framework which incorporates all these concepts. The entire *pattern* of service arrangements, service recipients, and service types, for any city, can be displayed in the form of a matrix, as shown in Table 7. Minor variations in this matrix would allow for the different types of service (noncombustibles, aluminum cans, and so on) which may apply to different communities.

The use of such a matrix facilitates comparisons between cities and permits a full but clear portrayal of a city's entire, complex pattern of collection services for municipal solid waste. As an illustration, the entries in Table 7 display the service pattern for the city of Indianapolis.

THE USE OF ALTERNATIVE COLLECTION ARRANGEMENTS

Having constructed this framework for classifying alternative organizational arrangements, the next step is to measure the prevalence of the different arrangements. Such measurement is not unprecedented. Between 1902 and 1974 no fewer than eleven major surveys examined the organization of refuse collection in American cities. However, these surveys, in the aggregate, suffered from a variety of defects which make it difficult to interpret the findings, to compare them, or to detect changes in the utilization of different organizational arrangements. The principal defects of prior surveys, taken as a group, were the following:

- –absence of a clear and coherent framework defining the organizational arrangements;
- –low response rates to survey questionnaires, with unknown biases in the returns;
- –use of mail surveys in all but one case, with ambiguous responses;
- –inclusion or exclusion of various classes of service recipients;
- –inclusion or exclusion of various classes of solid waste;
- –varying limits on the size of cities included in the surveys; and
- –presentation of data in ways that defy comparison.

THE COLUMBIA UNIVERSITY SURVEY

A survey, designed with a conscious effort to avoid the defects of the prior surveys, was conducted in 1975. It was carried out by telephone, instead of mail, to achieve the following:

TABLE 7
PATTERN OF MUNICIPAL SOLID WASTE COLLECTION SERVICES FOR A CITY[a]

	Service Recipient									
	Residential							Commercial	Institutional	Streets and Public Property
	Multiple Dwellings	Other Residential Dwellings								
		Mixed Residential Refuse (%)	Separated							
Service Arrangement			Wet Garbage	Yard Trash	Bulk Waste	News-papers	Other			
Municipal										100%
Contract										
Franchise										
Private	X	18			X			X	X	
Self-Service		5			X			X	X	
Special District	X	70			X					
Special District—Contract		7								
Intergovernmental Contract										
Association Contract										
Other										

a. Indianapolis, Indiana.

NOTE: Except as indicated, reliable estimates are not available for extent of use of the various arrangements.

(1) speed of completion of the survey;

(2) high response rate, thereby minimizing error due to bias in responses;

(3) accuracy, through verbal discussion and clarification where necessary; and

(4) firsthand impressions by interviewers experienced in local government as to the reliability of the data.

The primary basis for the inclusion or noninclusion of a given governmental jurisdiction in the sample was location in any Standard Metropolitan Statistical Area (SMSA) that lies entirely within a single state and has a total population less than 1,500,000 (these restrictions were imposed by the sponsor of the research). Thus, 200 of the 243 SMSAs (according to the 1970 Census) were eligible. Within these SMSAs, all incorporated municipalities with a population greater than 2,500 were eligible. In addition, townships were eligible if they were authorized by their state to collect or arrange for the collection of solid waste, when the population for which the township had such authority was 2,500 or greater. The total number of such governments is 2,060, and these communities constitute the statistical universe for the survey. Of these, 252 are central cities in the SMSAs, and 1,808 are satellite cities or townships. All eligible communities in 41 of the 200 eligible SMSAs were included in the sample. From the remaining 159 SMSAs, all central cities but only one-half of the satellite cities were included in the sample. The latter were then double counted to create the complete data file upon which the analysis is based. In summary, the total sample size consisted of 1,377 communities, the response rate was 99.9% (one community could not be reached by telephone despite repeated attempts), and, because of the nature of the statistical sample, it can be stated with confidence that the results describe the entire universe of 2,060 communities. It is this universe that is analyzed and described below. (The terms "city" and "community" are used interchangeably here and should be understood to mean both the cities and the townships described above.)

Because of the size of the sample, the way the sample was selected, the high response rate, and the fact that the survey was conducted by telephone, the author considers the findings presented here to be the most authoritative to date.

Altogether, the 200 SMSAs in the sample have a total population of 67.5 million, or 33.3% of the 1970 U.S. population; the 2,060 communities in turn have a population of 52 million. The sample is biased, by design, toward central cities and suburbs and ignores rural areas and small communities of less than 2,500 population. This bias reflects the study's focus on policy: refuse collection is a more significant problem in urban areas than in rural areas.

SURVEY FINDINGS

In the terminology introduced above, Table 8 shows the distribution of cities by service provider for each type of service recipient. Totals are over 100% in most cases because in some cities there are available (though not necessarily to a given recipient) more than one kind of service provider.

Service Delivery in Cities. It is clear that collection by private firms is quite commonplace for residential, commercial, institutional, and industrial service. Indeed, the number of cities that have private firms collecting at least some of their residential refuse is almost twice as large (66.7%) as the number that have municipal agencies collecting at least some of their residential refuse (37.4%). For commercial, institutional, and industrial refuse, cities with (at least

TABLE 8
PERCENTAGE OF CITIES IN WHICH A GIVEN SERVICE DELIVERER IS FOUND, BY TYPE OF SERVICE RECIPIENT

Service Recipient	Total Number of Cities	Municipality		Private Firm		Self-Service		Other	
		No.	%	No.	%	No.	%	No.	%
Streets	1,830	1,463	77.9	59	3.2	NA[b]		314	17.2
Parks	1,735	1,347	77.6	188	10.8	NA		227	13.1
Litter baskets	1,663	1,140	68.6	188	11.3	NA		338	20.3
Residential									
Bulk	1,848	926	50.1	750	39.5	436	23.6	66	3.0
Mixed	2,052	768	37.4[a]	1,368	66.7	376	18.3	19	0.9
Multiple dwellings	1,744	646	37.0	1,242	71.2	18	1.0	17	1.0
Institutional	1,942	616	31.7	1,196	61.6	115	5.9	207	10.6
Commercial	2,010	628	31.2	1,698	84.5	85	4.2	39	1.9
Industrial	1,579	334	21.2	1,099	69.6	232	14.7	141	8.9

a. That is, 37.4% of 2,052 communities have government agencies which collect at least some of the mixed residential refuse.
b. NA = not applicable.

some) collection by private firms outnumber cities with (at least some) municipal collection by two or three to one. Only for municipal properties—streets, parks, and litter baskets—do municipal agencies collect solid wastes in a significant majority of cities, as one would expect.

If one sets up a continuum between totally municipal collection and totally private collection, the various service recipients array themselves in a rather well-defined pattern. Municipal properties are served primarily by municipal collection, residences are served extensively both by municipal agencies and private firms, and the remaining categories—institutional, commercial, and industrial—are served predominantly by private firms. In these last three categories, service recipients tend to be served by private firms primarily because they require specialized service (for example, daily collection of large amounts of wastes) that the municipality, for various reasons, either cannot or will not provide.

The fact that the sample includes all cities with populations as small as 2,500, and does not include 15 of the 25 largest cities, partially explains the high proportion of cities with collection by private firms, as smaller cities are more likely to have private collection, and large cities are more likely to have collection by a municipal agency (this is discussed further below). Because of this fact, a rather different pattern emerges if one looks not at the number of *cities* serviced by private firms but at the number of *people* serviced by private firms.

Service Delivery by Population. One useful way to illustrate this is by weighting the figures in Table 8 by the population in each city. Table 9 does this. It shows the fraction of the population that lives in cities served by the indicated service provider, for each kind of service recipient. Again, where a city has both a municipal agency and one or more private firms providing service, that city's population is counted in both categories and the totals exceed 100%.

As expected, a comparison of Table 9 with Table 8 shows that because large cities are more likely to have municipal collection systems, the fraction of *population* which lives in communities with at least some municipal service is significantly higher than the fraction of *cities* with at least some municipal service, for almost every class of service recipient. For example, at least some of the mixed residential refuse is collected by the municipality in cities where 65.4% of the population lives, by private firms where 47.0% lives, and by self-service where 11.2% lives.

TABLE 9
PERCENTAGE OF POPULATION LIVING IN CITIES THAT HAVE THE INDICATED SERVICE PROVIDER, BY TYPE OF SERVICE RECIPIENT

	Service Provider			
Service Recipient	Municipality	Private Firm	Self-Service	Other
Streets	90.0	1.5	NA	6.1
Parks	81.8	4.2	NA	11.0
Litter baskets	30.3	5.0	NA	7.4
Residential				
Bulk	57.3	29.1	17.1	5.0
Mixed household	65.4[a]	47.0	11.2	1.4
Multiple dwellings	56.8	64.7	4.5	1.8
Institutional	47.9	59.6	7.3	8.6
Commercial	52.5	83.5	4.7	1.8
Industrial	24.7	70.3	14.2	5.9

a. That is, 65.4% of the population in the sample live in a city or town where some municipal workers collect at least some of the residential mixed refuse.

NA = not applicable.

Table 10 shows a third useful way to present the findings. Here the original figures from Table 8 are weighted both by population in each given city and by the percentage of that population that is served by the indicated provider—that is, the figures shown are absolute percentages of the total population. For the collection of mixed refuse from residences other than multiple dwellings, 61.3% of the population receives municipal service, 36.0% is served by private firms, and 1.2% practices self-service. These figures must be used with caution, however; the data for the fraction of the population in a community serviced by a given arrangement may be biased inasmuch as the information was obtained from municipal employees whose estimates of the magnitude of private-sector operations may be in error. (A prior study of the private sector concluded that 50% of the population receives service from private firms [Applied Management Sciences, 1973]).

TABLE 10
PERCENTAGE OF POPULATION SERVED BY INDICATED SERVICE PROVIDER, FOR COLLECTION OF MIXED RESIDENTIAL REFUSE FROM OTHER THAN MULTIPLE DWELLINGS

Service Provider			
Municipality	Private Firm	Self-Service	Other
61.3	36.0	1.2	1.5

Service Arrangements. Table 11 adds the concept of "service arranger" and shows the arrangements used in these cities. This table is, in some entries, identical to Table 8, as the "municipal" and "self-service" arrangements remain unchanged. However, where the service deliverer is a private firm the arrangement may be "contract," "franchise," or "private"; these, and other arrangements defined above, are shown.

Municipal properties–streets, parks, and litter baskets–are serviced by contract in 3% to 9% of cities. Mixed residential refuse is collected almost equally by municipal and private arrangements (approximately 38% of the cities use each), while 20.5% of cities award contracts and 8% award franchises for this service. Contract arrangements are more likely to be used for residential collection than for other types of collection. Self-service is used primarily for residential refuse and by industrial establishments.

Table 12 shows arrangement type by service recipient for those cities in which a given recipient is served *only* by one arrangement. This revealing table shows that, with respect to the collection of residential mixed refuse, 78.3% of the cities have only a single service arrangement, and *substantially more communities rely entirely on private firms (contract, franchise, and private–total 45.2%) than on municipal agencies (32.5%) for this vital service.*

Demographic Effects. Just as different service recipients are associated with different arrangements, various demographic factors are also associated with different arrangements. Table 13 shows that cities with larger populations are considerably more likely to have municipal service for residential collection and also that the larger the city, the more likely this is to be the case.

When the data are examined by region, some sharp differences emerge. Municipal collection predominates in the South by a wide margin and is far more likely to be found there than elsewhere. Franchise collection is much more common in the West than in other parts of the country, although it is not a dominant arrangement in any region.

The different forms of local government are quite similar in their use of the various collection arrangements, except for the relatively rare town-meeting form of government.

Specialized Residential Services. There are other, more specialized types of residential service that bear mentioning. Table 14 shows, for

TABLE 11
NUMBER AND PERCENTAGE OF CITIES, BY ARRANGEMENT, FOR EACH TYPE OF SERVICE RECIPIENT

Service Recipient	Total No. of Cities Reporting Service	Total Arrangements (sum of rows)		Municipal		Contract		Franchise		Private		Self-Service	
		No.	%	No.	%	No.	%	No.	%	No.	%	No.	%
Streets	1,830	1,836	100.3	1,463	79.9	55	3.0	–	NA	4	0.3	NA	
Parks	1,735	1,762	101.6	1,347	77.6	162	9.3	–	NA	26	1.5	NA	
Litter baskets	1,663	1,666	100.1	1,140	68.6	148	8.9	–	NA	40	2.4	NA	
Residential													
Bulk	1,848	2,178	117.9	926	50.1	190	10.3	47	2.5	513	27.8	436	23.6
Mixed refuse	2,052	2,531	123.3	768	37.4	421	20.5	165	8.0	782	38.1	376	18.3
Multiple dwellings	1,744	1,923	110.3	646	37.0	275	15.8	129	7.4	838	48.1	18	1.0
Institutional	1,942	2,134	109.9	616	31.7	231	11.9	111	5.7	854	44.0	115	5.9
Commercial	2,010	2,450	121.9	628	31.2	242	12.0	139	6.9	1,317	65.5	85	4.2
Industrial	1,579	1,806	114.4	334	21.2	109	6.9	93	5.9	897	56.8	232	14.7

Service Recipient	Special District–Contract		Inter-governmental Contract		County		Municipality–Special District		Voluntary Association		Special District		Other	
	No.	%	No.	%	No.	%	No.	%	No.	%	No.	%	No.	%
Streets	0	0	41	2.2	15	0.8	9	0.5	0	0	3	0.2	246	13.4
Parks	1	0.0	17	1.0	41	1.8	42	7.4	7	0.4	9	0.5	110	6.3
Litter baskets	0	0	17	1.0	9	0.5	8	0.5	9	0.5	4	0.2	291	17.5
Residential														
Bulk	4	0.2	1	0.0	2	0.1	0	0	8	0.4	0	0	51	2.8
Mixed refuse	6	0.3	4	0.2	2	0.1	1	0.0	0	0	0	0	6	0.3
Multiple dwellings	3	0.2	2	0.1	2	0.1	0	0	0	0	0	0	10	0.6
Institutional	3	0.2	5	0.3	8	0.4	3	0.2	0	0	1	0.0	187	9.6
Commercial	1	0.0	0	0	2	0.1	0	0	0	0	0	0	36	1.3
Industrial	1	0.0	0	0	2	0.1	0	0	0	0	0	0	138	8.7

NA = not applicable.

TABLE 12
NUMBER AND PERCENTAGE OF CITIES IN WHICH A CLASS OF SERVICE RECIPIENT IS SERVED BY ONLY A SINGLE ARRANGEMENT

Service Recipient	Arrangement											
	Total		Municipal		Contract		Franchise		Private		Self-Service	
	No.	%	No.	%	No.	%	No.	%	No.	%	No.	%
Streets	1,513	73.7	1,458	71.0	51	2.5	0		4	0.2	NA	
Parks	1,299	63.3	1,123	54.7	136	6.6	0		40	1.9	NA	
Litter baskets	1,306	63.6	1,124	54.7	142	6.9	0		40	1.9	NA	
Residential												
Bulk	1,523	74.2	787	38.3	170	8.3	35	1.7	297	14.5	234	11.4
Mixed	1,607	78.3	668	32.5	372	18.1	121	5.9	436	21.2	10	0.5
Multiple dwellings	1,578	76.9	514	25.0	250	12.2	122	5.9	682	33.2	10	0.5
Institutional	1,586	77.3	503	24.5	203	9.9	106	5.2	710	34.6	64	3.1
Commercial	1,603	78.1	369	18.0	152	7.4	120	5.8	942	45.9	70	1.0
Industrial	1,244	60.6	230	11.2	78	3.8	82	4.0	740	36.0	114	5.6

NOTE: Percentages are calculated on the basis of 2,052 cities.

NA = not applicable.

TABLE 13
SERVICE ARRANGEMENTS FOR COLLECTION OF RESIDENTIAL MIXED REFUSE

	Total	Municipal		Contract		Franchise		Private		Self-Service		Other	
		No.	%	No.	%	No.	%	No.	%	No.	%	No.	%
Total	2,531	768	30.3	421	16.6	165	6.5	782	30.9	782	14.9	19	0.8
Population Group	2,531												
>500,000	11	8	72.7	2	18.2	0		1	9.1	0		0	
250,000-500,000	26	19	73.1	2	7.7	0		3	11.5	1	3.8	1	3.8
100,000-249,999	96	62	64.6	8	8.3	2	2.1	15	15.6	7	7.3	2	2.1
50,000- 99,999	172	87	50.6	17	9.9	20	11.6	26	15.1	21	12.2	1	0.6
25,000- 49,999	203	63	31.0	35	17.2	25	12.3	53	26.1	27	13.3	0	
10,000- 24,999	503	179	35.6	117	23.3	34	6.8	117	23.3	54	10.7	2	0.4
5,000- 9,999	640	178	27.8	105	16.4	29	4.5	219	34.2	107	16.7	2	0.3
2,500- 4,999	880	172	19.6	134	15.3	56	6.4	348	39.5	159	18.0	11	1.2
Geographic Region	2,531												
Northeast	981	186	19.0	213	21.7	22	2.2	382	38.9	176	17.9	2	0.2
North Central	715	143	20.0	111	15.5	16	2.2	330	46.2	107	15.0	8	1.1
South	469	341	72.7	28	6.0	34	7.2	33	7.0	27	5.8	6	1.3
West	366	98	26.8	69	18.9	93	25.4	37	10.1	66	18.0	3	0.8
Metro/City Type	2,531												
Central	307	192	62.5	30	9.8	14	4.6	43	14.0	23	7.5	3	1.0
Suburban	2,224	576	25.9	391	17.6	151	6.8	739	33.2	353	15.9	14	0.6
Form of Government	1,799												
Mayor-council	876	374	42.7	214	24.4	42	4.8	178	20.3	64	7.3	4	0.4
Council-manager	724	319	44.1	109	15.1	103	14.2	100	13.8	87	12.0	6	0.8
Commission	67	34	50.7	14	20.9	3	4.5	9	13.4	4	6.0	3	4.5
Town meeting	112	16	14.3	16	14.3	2	1.8	43	38.4	35	31.3	0	
Representative town meeting	20	4	20.0	2	10.0	0		8	40.0	6	30.0	0	

NOTE: This table shows the distribution of *arrangements,* not the distribution of *cities.* The total number of arrangements is 2,531 in the 2,060 cities.

TABLE 14
ARRANGEMENTS FOR SPECIALIZED TYPES OF SERVICE

	Service Type			
	Wet Garbage	Leaves	Yard Trash	Newspapers
Total	128	895	464	45
Municipal				
No.	67	659	355	30
%	52.3	73.6	76.5	66.7
Contract				
No.	41	70	38	12
%	32.0	7.8	8.2	26.7
Franchise				
No.	4	35	20	1
%	3.0	3.9	4.3	2.2
Private				
No.	18	66	21	2
%	14.1	7.4	4.5	4.4
Self-Service				
No.	7	111	22	
%	5.5	12.4	4.7	
Special district				
No.		1		
%		0.1		
Special district—contract				
No.		2		
%		0.2		
Intergovernmental—contract				
No.		2	1	
%		0.2	0.2	
Special district—municipal				
No.		2		
%		0.2		
Other				
No.		2	7	
%		0.2	1.5	

instance, that in 128 cities (6.2% of the sample) wet garbage and dry rubbish are collected separately. Separate collection of leaves occurs in 895 cities (43.4% of the sample), although in 111 of these cities the service recipient must dispose of them himself. Yard trash is collected separately in 464 cities (22.5% of the sample) and newspapers in 45 cities (2.2% of the sample). The latter figure probably understates the extent to which newspapers are collected separately, insofar as voluntary recycling efforts may be under-reported if such collection does not fall under the purview of the agency or firm concerned with the collection of most other residential solid wastes. The arrangements utilized for these separate collection services also appear in Table 14.

SOLID WASTE DISPOSAL

Organizational arrangements for solid waste disposal can be classified in somewhat the same manner as the arrangements for solid waste collection. Instead of the community or the city being the unit of analysis, however, the individual disposal site or facility is the basic unit of analysis. Disposal sites may be classified in terms of four parameters:

(a) type of disposal facility
(b) ownership
(c) operation
(d) type of waste accepted.

TYPE OF FACILITY

In this study, seven different types of disposal sites are identified: dumps, modified landfills, sanitary landfills, transfer stations, incinerators, resource-recovery facilities, and refuse processing facilities (e.g., shredding, baling).

OWNERSHIP

Disposal sites are classified as to their owners. They may be owned by a government or by a private party.

OPERATION

A waste disposal facility is operated on a day-to-day basis by the owner, by a private party, or—when the owner is a government jurisdiction—by another government jurisdiction.

TYPE OF WASTE ACCEPTED

Many disposal sites restrict the kinds of waste materials they will accept; therefore, it was reasonable to classify disposal sites in terms of such restrictions. Six major kinds of facilities could be identified: those that accepted (1) any kind of waste; (2) primarily residential and commercial waste; (3) only nonputrescible waste; (4) only construction and demolition waste; (5) only nonhazardous wastes; and (6) any other restricted class of material.

OWNERSHIP AND OPERATION OF DISPOSAL SITES

A useful representation of the organizational arrangements for disposal can be obtained by looking at the pattern of ownership and operation of the facility and by ignoring for the moment the type of facility and the kinds of wastes it accepts. Table 15 shows this pattern for the waste disposal sites in the 281 counties of the 200 SMSAs referred to above.

More than half the sites (55.6%) are owned and operated by government agencies, while almost one-third (31.1%) are owned and operated privately. Government ownership with private operation is quite rare (only 2.7% of the sites) and is actually less common than private ownership and government operation (7.5% of the sites); in the latter cases, typically, government leases the land from a private owner and uses it as a landfill site.

SUMMARY

The demand for greater productivity in local government requires that attention be paid to a most basic issue concerning municipal services: who can best perform them?—that is, what organizational structure is most efficient and effective in delivering local government services? In examining this issue, the starting point is to create a conceptual framework for classifying the alternative organizational arrangements, and to measure the extent to which each is used. This was done for two specific services: refuse collection and refuse disposal.

In order to describe the organization of collection services for municipal solid waste in a community, it is necessary to identify the

TABLE 15
DISPOSAL SITES, BY TYPE OF OWNER AND OPERATOR

	Operator							
Owner	Private		Government		Other Government		Total	
	No.	%	No.	%	No.	%	No.	%
Private	196	31.1	47	7.5	NA	NA	243	38.6
Government	17	2.7	350	55.6	19	3.0	386	61.3
Totals	213	33.8	397	63.1	19	3.0	629	99.9

NA = not applicable.

kind of organization which *arranges* for service and the kind of organization which *provides* the service to each class of *service recipient,* for each *type of solid waste* that is collected separately. To this information must be added information on the *fraction* (or number) of service recipients who are served by each arrangement. A number of organizational *arrangements,* that is, pairs of service arrangers and service providers, are identified. They can be utilized, singly or in combination, to supply the various types of service to the different classes of service recipients, thereby creating a *pattern* of arrangements for solid waste collection services in a city.

For refuse disposal, service arrangements are defined according to the type of facility, the owner of the disposal facility, the operator, and the restrictions on the type of waste accepted.

A study was conducted of cities in metropolitan areas of the United States with populations exceeding 2,500; 2,060 cities with a combined population of 52 million were included. It was found that private firms collect commercial, institutional, and industrial refuse in about three times as many cities as municipal agencies do, and they collect residential refuse in almost twice as many cities (66.7%) as municipal agencies do (37.4%). Substantially more communities (45.2% of all cities) rely entirely on private firms for collection of mixed residential refuse than on municipal government (32.5% of all cities). However, because large cities tend to utilize municipal collection for residences, most of the people (61.3%) are serviced by municipal agencies, compared to 36.0% serviced by private firms.

Municipal, contract, franchise, private, and self-service arrangements account for more than 99% of all arrangements, although sixteen different arrangements in total were found. The number of cities which utilize each arrangement to collect mixed residential refuse from at least some of their residents is as follows: municipal: 37.4%; contract: 20.5%; franchise: 8.0%; private: 38.1%; self-service: 18.3% (the total exceeds 100% because in some cities more than one collection arrangement is utilized). Municipal collection is particularly common in Southern cities; franchise collection is more popular in the West than elsewhere.

With respect to refuse disposal sites, 55.6% are owned and operated by local governments, while 31.1% are owned and operated by private firms.

In short, a wide variety of organizational arrangements is utilized to provide refuse collection and disposal services for urban dwellers. Because public agencies and private firms play prominent roles in the provision of these important services, the field of refuse collection and disposal affords an excellent opportunity to evaluate the relative performance of the public and private sectors.

REFERENCES

American Public Works Assocation [APWA] (1975) Solid Waste Collection Practices. Chicago: Public Administration Service.

--- (1966) Refuse Collection Practice. Chicago: Public Administration Service.

Applied Management Sciences (1973) The Private Sector in Solid Waste Management–A Profile of Its Resources and Contribution to Collection and Disposal. Washington, D.C.: U.S. Environmental Protection Agency.

Council of State Governments (1973) The States' Role in Solid Waste Management. Lexington, Kentucky.

Council on Environmental Quality (1970) Environmental Quality: First Annual Report. Washington, D.C.: Government Printing Office.

DOWNS, A. (1970) Urban Problems and Prospects. Chicago: Markham.

DRUCKER, P. F. (1968) The Age of Discontinuity. New York: Harper & Row.

FITCH, L. C. (1974) "Increasing the role of the private sector in providing public services," in W. D. Hawley and D. Rogers (eds.) Improving the Quality of Urban Management. Beverly Hills, Calif.: Sage.

International City Management Association (1964) Contracting for Municipal Services. Management Information Service Report 240. Washington, D.C.: I.C.M.A.

National League of Cities (1974) America's Mayors and Councilmen: Their Problems and Frustrations. Washington, D.C.: National League of Cities.

NISKANEN, W. A. (1971) Bureaucracy and Representative Government. Chicago: Aldine.

Public Health Service (1968) 1968 National Survey of Community Solid Waste Practices: Preliminary Data Analysis. Washington, D.C.: U.S. Department of Health, Education and Welfare.

SAVAS, E. S. (1974) "Municipal monopolies versus competition in delivering urban services," in W. D. Hawley and D. Rogers (eds.) Improving the Quality of Urban Management. Beverly Hills, Calif.: Sage.

--- (1971) "Municipal monopoly." Harper's Magazine (December): 55-60.

SMITH, F. A. (1975) Comparative Estimates of Post-Consumer Solid Waste. Report SW-148. Washington, D.C.: U.S. Environmental Protection Agency.

9

Municipal Fire Protection Performance in Urban Areas: Environmental and Organizational Influences on Effectiveness and Productivity Measures

PHILIP B. COULTER
LOIS MacGILLIVRAY
WILLIAM EDWARD VICKERY

□ THE PRESIDENT'S COMMISSION on Fire Prevention and Control reported an enormous annual loss of life and property stemming from uncontrolled fire. Each year fires occasion approximately 12,000 deaths and an estimated cost of $11.4 billion (including property loss, fire department operation, burn injury treatment, insurance industry operating cost, and productivity loss). It is clear that greater effort is needed to improve the effectiveness and efficiency of public programs designed to prevent and suppress fires. However, before programs can be redesigned to reduce the magnitude of human suffering and economic loss, a greater understanding of the urban fire service organizational infrastructure and its correlates is essential.

AUTHORS' NOTE: *The findings reported here are taken from an ongoing four-year research program in evaluation of the organization of fire service delivery currently in progress in the Center for Population and Urban-Rural Studies (CPURS), Research Triangle Institute. This research program was supported by the Division of Research Applied to National Needs (RANN) in the National Science Foundation. The authors wish to thank Sally Plotecia, an economist in CPURS, for a very thoughtful critique of an earlier version of this manuscript.*

This article was prepared with the support of NSF Contract C-900. However, any opinions, findings, conclusions or recommendations herein are those of the authors and do not necessarily reflect the views of the National Science Foundation.

The research reported here is part of a larger project[1] still in progress, one important aim of which is to investigate the organizational features of municipal fire service delivery systems and to evaluate alternative organizational arrangements for providing fire protection. Although various criteria are available for such evaluation, only effectiveness and efficiency are employed here.[2]

We begin by offering a statement of the problem, followed by a brief summary of the relevant literature on fire service organization and delivery. We then proceed to some substantive and operational definitions of key concepts, and the statement of three general hypotheses to be tested. Next we describe the data base and the methodology used to test the hypotheses. Finally, we present the results of the analysis and discuss their policy implications.

I. THE PROBLEM

The definition of the fire protection service problem must begin with the metropolitan area in which dense gatherings of population and diverse activities dictate the form, consequence, and variety of publicly provided goods and services. A service delivery system must respond to the characteristics of the metropolitan area; it cannot reorganize or modify them in a major way. Yet it still must satisfy certain basic objectives. For example, the responsibility of the fire protection service is to prevent uncontrolled fires, to minimize the human and property loss in the event of ignition, and to confine and suppress a fire once ignited. It follows, however, that the requirements of this service are not uniformly distributed within a given space. This implies that greater protection is afforded to the more combustible and the more critical or valued areas (using either economic or social criteria) and that the major dimensions of variation in such protection warrant study in relation to variations in the demand for these services by the population and activities characteristic of the area.

For our purposes, fire protection shall be defined as consisting of two aspects: fire prevention and fire control. Fire prevention in the largest sense involves the interplay of land-use planning, architectural design, building material and furnishing characteristics, and individual care in handling flammable materials, tools, and so on. Although most of these activities are performed in the private sector, the land values, insurance rates, and neighborhood safety of all citizens are affected by their performance. Generally, then, an oversight, regulatory, or educational function is organized in the public sector for the commonweal. Building fire safety and related inspection programs provide examples of these municipal oversight and regulatory functions. Fire control or suppression involves detection of fire, alarm dispatch response by the firefighters, containment, and, finally, suppression of the fire.

The objectives are to make both prevention and suppression of fires more effective (reduce the incidence of fires and minimize the losses in those that do occur) and more productive (minimize the expenditures for any given level of effectiveness). Achieving such objectives presumes a clear understanding of the organizational and behavioral characteristics of fire service delivery within the constraints imposed by characteristics of the community. That is, the possibility of altering an existing performance level depends on an understanding of those aspects of the situation that can change and those that can only be controlled. The small but growing body of literature on fire service delivery provides some assistance in identifying major fire protection service dimensions and in framing hypotheses to test their interrelationships. It is to a brief critique of that literature that we now turn.

II. REVIEW OF THE LITERATURE

Previous research on fire service delivery has been concerned primarily with two topics and their variants: budget expenditures for fire service or "output" (cost or supply functions) and demand or need for service. In the case of research on total expenditures or policy output, the principal efforts have been (1) to identify and measure the influence of variables which affect fire service expenditures and (2) to determine if economies of scale exist and, if so, at approximately what range of population size. The body of literature on demand for service is much smaller. It focuses on estimating the influence of factors assumed to affect the incidence of fire or the magnitude of property loss due to fire within a municipality.

With few exceptions, previous research on fire service expenditures has operationally defined the output measure as total annual fire department budget expenditure per capita (Bahl, 1969; Bahl and Sanders, 1965; Brazer, 1959; Fisher, 1964; Hirsch, 1959; Masotti and Bowen, 1965; Pidot, 1969; Pulsipher and Weatherly, 1968; Sacks and

Harris, 1964; Schmandt and Stephens, 1960). Typically, this dependent variable was regressed on a series of variables chosen to explain as much of the per capita expenditure variation as possible and to determine the independent variables that are most closely related to fire service output per capita. The three most frequently used and probably most influential are per capita income, population density, and percent urban residence (Brazer, 1959; Bahl, 1969; Bahl and Sanders, 1965). In combination, they account for approximately 60% to 85% of the variation in per capita fire expenditure, depending on the data base used and the year or period of observation. When other variables[3] are included in the analysis, the percentage of fire service output variation explained increases only marginally (Bahl, 1969). Of these additional variables, however, three are clearly more important for fire service expenditures: ratio of property taxes to total general revenues, ratio of intergovernmental revenues to total general revenues, and per capita retail sales (Bahl, 1969). Different results are achieved with different specifications of the dependent variable, expenditures[4] (Hitzhusen, 1972). Finally, Czamanski (1975) developed a model which emphasizes an inverse relationship between fire losses and expenditures on fire protection; his model suggests that public decision makers are faced with the problem of minimizing the sum of property losses and expenditures on fire protection services. Their optimum point of "preventive efficiency" in this model, therefore, is the point at which the change in total losses with respect to the change in fire loss just equals the change in total losses with respect to the change in fire protection expenditures. In other words the optimum expenditure level is that point where the slope of the cost function equals unity.

The other major concern of research on cost has dealt with economies of scale. Studies of possible scalar economies for the cost of fire service delivery have produced mixed and, to some extent, conflicting results (Will, 1965; Hirsch, 1970; Ahlbrandt, 1973; Hitzhusen, 1972).

With regard to a curve which accurately and consistently describes the relationship between total population size and per capita cost of fire service, there is little specific agreement on either the shape of the curve or the dollar and size or area value of the minimum and maximum. There does seem, however, to be some agreement that there may be appreciable economies of scale up to a population of about 10,000, some limited or no economies of scale from about 10,000 to 100,000, and diseconomies or no economies of scale from about 100,000 to about 300,000. Similarly it was found that there are economies of scale as fire service output expands and the number of volunteers increases until, for fully paid departments, output expands at a constant cost per capita, i.e., no further economies of scale (Ahlbrandt, 1973). Another study provides some evidence that there are economies of scale as the number of fire service subfunctions performed increases (Schmandt and Stephens, 1960).

The purpose of developing a fire service demand function is similar to that of the expenditure function, i.e., to predict and explain variations in demand, to identify variables most closely associated with demand, and therefore, presumably, to learn something about how to respond to or control demand in terms of certain public safety goals or standards. There has been much less research published on fire service demand, but in all known cases, demand is measured either as number of alarms of various types or a fire loss index such as property damage per sales value of property. This research has also been limited to intracity, i.e., neighborhood, variations in demand. Variation in the number of alarms is at least partially a function of time of day (more frequent alarms in late afternoon), demography (more frequent alarms in more densely populated parts of the city, and weather (fewer alarms during periods of rainfall (Chaiken and Rolph, 1970). The demand function, which uses property loss per sales value of property, produced a coefficient of determination of 95% by employing eight social, economic, and physical variables similar to those used in regression analyses of cost functions (Syron, 1972). However, the three variables which proved to be significantly and strongly related to the fire loss index were race (percentage nonwhite), education (percentage of adults with greater than eighth-grade education), and vacant housing (percentage of houses vacant).

Many of these research findings stand in contradiction to one another, yet, none of the work truly replicates any of the others. Each scholar specified his equations somewhat differently, making it difficult to weigh the relative merits of each. The samples used exacerbate the difficulty in evaluating this body of literature in that not one is a sample interpretable much beyond the study universe. Thus it is entirely possible that the differences in findings concerning scale economies are artifacts of sampling error. Moreover, because the authors did not provide conceptual frameworks for choosing municipal or organizational characteristics used as explanatory variables, the application of findings is limited.

While the conceptual, methodological, and mensurative deficiencies of this literature severely limit its usefulness, it does supply answers to the question "What social and economic characteristics of communities are related to their fire service spending patterns?" There is, however, little attention to *how* or *why* these variables are related, i.e., how the processes operate by which service demand is translated into service delivered and population affected. No attention whatsoever was given to the mechanisms by which fire service dollars and organization are "delivered" to the community, i.e., no organizational, behavioral, or legal variables were brought to bear on the question.

III. CONCEPTUAL FRAMEWORK

Despite the shortcomings of previous research, it does suggest the relevance of certain explanatory variables for hypotheses about effectiveness and productivity of fire service delivery. Effectiveness and productivity (Shaenman and Schwartz, 1974; International City Management Association, 1974) are proposed as the main criteria for the evaluation of fire service delivery. Effectiveness is defined as the extent to which the fire service avoids or reduces property loss, death, and injury due to fire. More specifically, both prevention and suppression can be evaluated with respect to their effectiveness. Prevention effectiveness refers to the degree to which the fire service avoids or minimizes the incidence of fire. Suppression effectiveness refers to the extent to which the fire service minimizes loss per incident. Such ratios as the number of incidents per 1,000 population and dollar property loss per capita would be measures of prevention effectiveness and suppression effectiveness, respectively. It is well known that a service delivery system can be quite effective in achieving its objectives, but at an unreasonably high cost, and therefore is inefficient. Similarly, a relatively ineffective fire service can be relatively efficient. Productivity, on the other hand, measures the cost of varying levels of effectiveness. Effectiveness and efficiency are theoretically independent. A measure of productivity combines effectiveness and efficiency.

A ratio which measures this concept of productivity is total cost of fire (budget expenditure plus loss) per capita. Presumably, there is an inverse relationship between expenditure and loss; as expenditure increases, loss declines. Thus the total cost of fire includes dollars spent for prevention and suppression *and* property loss in dollars. The sum of these cost components is to be minimized. Such a productivity concept captures variations in both the effectiveness and efficiency of both prevention and suppression.

Based on these definitions, Table 1 summarizes four important measures of service delivery or performance that are to be evaluated here: prevention effectiveness, suppression effectiveness, expenditures and fire service productivity.

All of the performance measures in Table 1 are to be minimized. The lower the number of alarms per 1,000 population in a given municipality, the more effective is the prevention program of its fire service. The smaller the property loss per 1,000 population, the more effective is the municipality's fire suppression performance. The objective here is to identify a set of variables which will help to explain why different cities are observed to have different levels of performance. The literature reviewed earlier provides numerous suggestions about relevant socioeconomic, physical, and climate variables. There is, however, very little in the previous literature that suggests the importance of organizational variables in general or that empirically demonstrates the relevance of particular features of organization *for effectiveness or productivity*. For the purpose of exploring the behavior of these aspects of performance, three general hypotheses are proposed:

(1) The supply of fire service reflects the demand for such service.

(2) The effectiveness of fire service delivery is dependent upon the organizational character of the service.

(3) The productivity of fire service delivery is dependent upon the organizational character of the service.

The first hypothesis simply states that variations in the magnitude and to some extent the kind of fire service provided by municipalities is a function of the variation in their urban environmental character or context. Such environmental characteristics determine

TABLE 1
PERFORMANCE MEASURES

1. Expenditures	Annual fire department budget per capita
2. Prevention effectiveness	Number of fires per 1,000 population
3. Suppression effectiveness	Dollars property loss from fires per 1,000 population
4. Productivity	Total cost (i.e., loss plus expenditure) per capita

to a large extent the nature of the fire hazard. The second and third hypotheses, while recognizing that effectiveness and productivity can be in part a function of urban environment as well as of technological capabilities, suggest that effectiveness and productivity are also influenced by the organizational and delivery characteristics

TABLE 2
ENVIRONMENTAL CONSTRAINTS ON FIRE SERVICE DELIVERY PERFORMANCE

Name	Operational Definition
I. Demographic	
A. Land area	Number of square miles protected
B. Density	Number of persons per square mile protected
C. Population size	Total number of residents
II. Housing quality	
A. Deterioration	Principal components method of factors analysis using orthogonal rotation with the varimax criterion was performed on 46 variables describing poor or fire-hazardous qualities of houses, their contents, and their residents. It produced this deterioration component clustering mainly old age and low value of physical housing structures. Factor scores on this component make up the housing deterioration variable.
B. Crowding	See II-A; crowding is the second component clustering mainly crowding conditions within homes for various kinds of residents.
III. Climate	
A. Very cold climate	Average annual number of days with temperature less than 0° Fahrenheit
B. Precipitation	Average annual precipitation
C. Windy conditions	Average January wind speed
D. Thunderstorms	Average annual number of thunderstorms
IV. Socioeconomic	
A. Property value	Tax revenues
B. Industrialism	Number of persons employed in manufacturing
C. Aged component of dependent population	Number of persons over 64 years of age living alone
D. Youth component of dependent population	Number of persons less than seven years of age
E. Social class	Number of persons with some college education
F. Institutionalized population	Number of persons in mental hospitals, homes for the aged and dependent, and other institutions
G. Water supply	ISO deficiency points for water supply
V. Governmental	
A. Council-manager form	Dummy variable in which council-manager form = 1 and all other forms = 0

of the fire service. Each of these hypotheses is specified and tested, and the results are presented in Section V (Findings).

The 17 city characteristics selected as relevant to variations in fire service performance are listed and operationally defined in Table 2. These 17 constraints can be categorized into the following five classes: demographic, housing quality, climate, socioeconomic, and governmental variables. Most of them are constraints that policy makers cannot readily change; they are not manipulable. Once their influence on various performance criteria is properly identified, however, policy makers and administrators can more rationally take their influence into account in making decisions about the more changeable aspects of the system. Table 3 lists and defines the variables that describe organizational features of fire service delivery systems, categorized according to fire suppression and fire prevention variables. There is some overlap (e.g., status of fire chief), but by and large each variable tends to fall into only one category.

IV. DATA COLLECTION AND METHODOLOGY

This conceptualization and measurement of the urban environment and fire service organization requires an extensive data collection effort to measure service delivery area characteristics and fire service organization characteristics, including records of alarms, expenditures, and fire losses. To satisfy these data requirements, surveys were carried out in a sample of 50 standard metropolitan statistical areas in the United States that have less than 1.5 million residents and are contained within a single state. Fire departments, building inspection departments, and city managers were surveyed by mail as part of a primary data collection effort for this project. These surveys were supplemented with data from the National Fire Protection Association's Survey of Public Fire Departments. In addition, extensive U.S. Census data on population, housing, and climate were gathered; Census of Governments data, information from the *City and County Data Book,* and summaries of the legal aspects of fire service were also collected.

Although data were collected for over 1,500 organized jurisdictions, which ranged in size from tiny to large, the analysis reported herein focuses on municipalities with a population of 25,000 or greater in which 87% of the population of these SMSAs resides. There were 324 such cities in our sample, although data are not complete for all cases.

TABLE 3
ORGANIZATIONAL VARIABLES WHICH AFFECT FIRE SERVICE DELIVERY PERFORMANCE

Name	Operational Definition
I. Fire Suppression	
A. Status of fire chief	Volunteer, part-time paid, full-time paid
B. Maximum response time	Longest response time in minutes to any part of jurisdiction with structures
C. Firefighter training	Standard work day allotment of time for skill maintenance and development drills in hours per month
D. Unionism	Based on whether any union represents a majority of fire personnel, which union, which ranks are represented, and the contents of any written labor agreements relative to fire personnel
E. Full-time paid fire department personnel	Number of fire department personnel who are full-time and paid employees
F. Fire service planning	Principal components method of factor analysis using orthogonal rotation and the varimax criterion was performed on 29 variables describing "organizational complexity" of the fire department. Fire service planning was the first component extracted; it includes such items as whether a department has a 5-10 year plan, and if so, does the plan include 11 specific elements. Factor scores on this component compose the fire service planning variable.
G. Administrative size	Total number of chiefs, assistant chiefs, deputy chiefs and battalion chiefs
H. Emergency response versatility	See I-F in this table. Emergency response versatility was second factor comprising items describing variability of response (e.g., number of fire fighters and pieces of apparatus) to various types of alarms (e.g., kind of fire, time of day, season of the year, and so on). Factor scores on these items are the response versatility variable.
I. Emergency rescue/ medical service (EMRS)	See I-F in this table. Component 3 clustered items describing whether a department provided emergency rescue, emergency medical care, emergency ambulance transportation, and nonemergency ambulance transportation. Factor scores on this item are the EMRS variable.
J. Further education incentive	Dummy variable: 1 point = departments that pay for fire fighters time off for education, 0 = those that do not
K. Constant manning	Dummy variable: 1 point = department that assigns a standard number of fire fighters per apparatus on each shift, 0 = those that do not
L. Mutual aid	Number of mutual aid calls responded to per year
M. Contract service	Number of alarms responded to in other jurisdictions with which department has a fire suppression service contract

TABLE 3 (continued)

Name	Operational Definition
II. Fire Prevention	
A. Building inspectors' training	Principal components method of factor analysis using orthogonal rotation and the varimax criterion was performed on 20 variables describing training, educational, and experience requirements for five types of inspectors. Component 1 is made up of items pertaining to training; factor scores on this factor comprise the building inspectors' training variable.
B. Building inspectors' education and experience	See II-A in this table. The second component grouped education and experience, factor scores of which make this variable.
C. Inspection program comprehensiveness	Score based on how many municipal departments (0-9) are involved in five phases of the building inspection process; possible scores vary from minimum of 0 to maximum of 45.
D. Inspection staff size	Number of municipal officials who actually perform any type of building and/or housing safety inspection
E. Fire fighter inspections	Number of fire safety inspections conducted by fire department personnel
F. Number of municipal inspection agencies	2 points = both fire department and any other municipality performing inspections, 1 point = either, 0 points = neither

This extensive data collection effort yielded sufficient data for the construction of all the variables reported in Tables 1-3. As stated earlier, the purpose is to account for variations in fire service effectiveness and productivity with both environmental and organizational variables. To this end, each performance measure listed in Table 1 (prevention effectiveness, for example) was categorized into quartiles. Cities were arrayed from lowest to highest score on prevention effectiveness and then divided into four equal-sized groups or quartiles. The same procedure was then followed in grouping municipalities with regard to their scores on the other three performance measures. This procedure was undertaken largely because two of the crucial performance indicators (fire incidence and fire loss) can experience radical variation from one year to the next.[5] Grouping cities into quartiles presumably eliminates the worst of such measurement instability.

The technique selected to evaluate performance is discriminant analysis. Discriminant analysis is a statistical method designed to distinguish two or more groups of cases by weighting and linearly combining variables expected to differentiate them so that the groups are forced to be as statistically distinct as possible (Nie et al.,

TABLE 4
MEANS OF SIGNIFICANT VARIABLES IN STEPWISE DISCRIMINANT ANALYSIS OF FIRE DEPARTMENT EXPENDITURES PER CAPITA

	Lowest Expenditure	Less Expenditure	Moderately High Expenditure	Highest Expenditure	All Cities	Wilks's Lambda[a]
1. Unionism score	11.7	12.0	17.7	23.7	17.8	0.80
2. Contract alarms	1152.12	11.03	4.86	40.65	101.4	0.70
3. Precipitation	722.12	624.34	699.69	881.95	744.02	0.63
4. Local alarms	564.87	4362.79	2473.80	4927.66	3669.80	0.58
5. Constant manning	0.50	0.79	0.89	0.87	0.83	0.53
6. Social class	0.26	0.31	0.27	0.21	0.26	0.51
7. Very cold climate	100.0	65.2	64.3	81.2	72.9	0.48
8. Density	3070.5	3828.0	3559.6	7274.3	4866.2	0.46
9. Emergency rescue/ medical service	0.32	–0.03	0.04	0.20	0.10	0.44
10. Mutual aid calls	34.5	29.1	27.5	10.9	22.7	0.43

a. All steps are significant at the .001 level or better.

1975). The objective is to discriminate as well as possible among the groups in terms of the discriminating variables.[6]

A second technique in discriminant analysis identifies the probable group membership of each case when the only information known about it is its values on the discriminating variables; the technique assigns each case to the group to which it "probably" belongs.[7] The percentage of correct classifications for each group and for all cases can then be computed. A stepwise entry of variables into the successive linear combinations was specified to order the discriminant variables on the basis of their discriminating power and statistical significance, measured by a reduction in the value of Wilks's Lambda which varies from 1.0 to 0.0. With the addition of each significant discriminant variable[8] into the equation, the value of Wilks's Lambda is reduced. The greater the reduction, the more important the variable(s). The value of Wilks's Lambda upon the entry of the last variable in the stepwise analysis indicates how much "indiscrimination" still remains among the four quartiles. The canonical correlation coefficients (Nie et al., 1975; Van de Geer, 1971) can be approximately interpreted as multiple correlation coefficients.

V. FINDINGS

This section reports and discusses the results of the application of discriminant analysis to each performance measure in Table 1 and to an appropriate set of discriminant variables from Tables 2 and 3. The order of discussion is: expenditures, prevention effectiveness, suppression effectiveness, and productivity.

FIRE DEPARTMENT EXPENDITURES

A large portion of municipal public expenditures for fire protection does not appear in the fire department—the costs of building inspection and water, for example. Yet, since the evidence suggests that the cities which spend more for suppression also spend more for prevention,[9] the fire department budget probably represents total expenditure rather well. Expenditure per capita was hypothesized to be sensitive to 20 variables—13 environmental features and seven organizational characteristics.[10] Only 10 of these 20 variables were actually able to contribute significantly to discrimination among four quartiles of budget expenditures per capita. Table 4 reports the mean average of each quartile on the 10 significant discriminant variables. The significant variables are listed from most to least discriminant.

Comparing the towns with the lowest and highest per capita fire department budget, the former are less unionized, have many more contract service alarms to answer, have less precipitation on the average, have fewer local alarms, tend to follow a more flexible manning practice, have higher social class, have more days below 0° Fahrenheit, are less dense, maintain a larger emergency rescue/medical service program, and have a greater number of mutual aid calls to which they respond. Five of these characteristics pertain to organizational structure and process and could, therefore, be changed with the possible result of altering the size of per capita expenditures. The relative importance of each discriminant variable is also detailed in Table 4.

The importance of unionism to high per capita budget expenditures is indicated by the 20-point reduction in Wilks's Lambda in Table 4. This variable discriminates well among all four quartiles of spending. The importance of unionism for expenditures (a positive relationship) may be to some extent a reflection of or surrogate for city population size because the number of local alarms is comparatively very low for the low-expenditure quartile, and number of alarms is very closely related to population. Larger cities have more alarms and also more unionized fire departments. Contract alarms, precipitation, and constant manning all contribute significantly and powerfully to discrimination among the four groups of cities. It seems that a very high number of responses to contract alarms coincides with a very handsome contract fee for service rendered. More precipitation tends, to some extent, to mean lower expenditures. Constant manning is expensive in terms of salaries, so towns that do not use it apparently spend less on salaries.

The first two discriminant functions are strong and significant in terms of differentiating the four quartiles of expenditure per person; the third function fails to pass the significance test. The first two canonical correlations are 0.60 and 0.51, respectively. Approximately 62% of the total variation in per capita budget is accounted for by these two significant linear combinations of discriminant variables. The explained variation is almost equally divided between the two functions. Finally, the linear functions were used to sort each city into its probable quartile; these predictions are incorporated in Table 5.

The linear functions do very well in correctly predicting the lowest

TABLE 5
PERCENTAGE OF CORRECT PREDICTIONS OF LEVEL OF PER CAPITA EXPENDITURE

	No. of Cases	Lowest Expenditure	Less Expenditure	Moderately High Expenditure	Highest Expenditure
Lowest expenditure	41	65.9	14.6	9.8	9.8
Less expenditure	41	7.3	58.5	31.7	2.4
Moderately high expenditure	41	7.3	31.7	41.5	19.5
Highest expenditure	41	4.9	4.9	26.8	63.4
Total	164				

Percentage of correctly predicted cases: 57.3%.

and highest per capita expenditure categories, fully 65.9% and 63.4%, respectively. Discrimination among the extremes is fairly successful. In fact, the 10 discriminant variables do quite well in predicting 58.5% of the less-expenditure cities correctly. It is the moderately high-spending cities for which the discriminant functions are least successful, and these bring the overall correct prediction rate down to 57.3%. In general, although all four quartiles of cities are different with regard to the discriminant variables, the two extreme quartiles and the less-expensive quartile are radically dissimilar.

FIRE PREVENTION EFFECTIVENESS

Ten environmental constraint variables and eight organizational variables were chosen to discriminate among quartiles of fire incidence per 1,000 population or prevention effectiveness.[11] The stepwise discriminate analysis, using three functions, eliminated all but 11 variables. The means of the 11 discriminant variables are presented in Table 6 in the order of their discriminating power. The most useful comparison of means is probably between the most effective and the least effective categories of cities.

An untutored examination of Table 6 might suggest that municipalities with the most effective prevention are most effective because, in comparison to the least effective group, they have less comprehensive inspection programs, less stringent education and experience requirements for their building inspectors, fewer full-time paid fire fighters, and a smaller inspection staff. What in fact seems

TABLE 6
MEANS OF SIGNIFICANT VARIABLES IN STEPWISE DISCRIMINANT ANALYSIS OF PREVENTION EFFECTIVENESS

	Most Effective	Moderately Effective	Less Effective	Least Effective	All Cities	Wilks's Lambda[a]
1. Inspection program comprehensiveness	3.72	8.55	9.94	5.10	7.65	0.89
2. Building inspectors' education and experience	0.50	0.48	1.25	0.77	0.81	0.79
3. Full-time paid fire department personnel	81.63	139.17	154.84	264.29	171.48	0.72
4. Land area (square miles)	113.53	50.40	54.67	44.29	56.25	0.66
5. Social class	30.5%	26.3%	29.5%	20.3%	26.2%	0.62
6. Fire fighter inspections	440.72	2971.0	3731.5	1723.4	2654.8	0.58
7. Industrialism	9331.8	8702.9	9731.0	14157.7	10525.7	0.55
8. Very cold climate	80.1	81.9	41.9	86.3	71.3	0.52
9. Inspection staff size	6.9	12.8	11.1	14.7	12.2	0.51
10. Property value (in 1,000s)	5794.2	3728.9	4616.8	9506.5	5727.1	0.47

a. All steps are significant at the .004 level or better.

to be happening is a clustering of higher class residential suburbs in the "most effective" category. They have larger land area, much higher social class, rather low level of industrialism, and a fairly high property value. In short, communities of this sort require considerably less prevention effort because their fire hazard and "natural" incidence rates are rather low; hence, they have a high effectiveness rating in spite of their low prevention effort.

Table 6 also reveals the actual contribution of each variable, in the three functions derived, to the discrimination among the four quartiles of prevention effectiveness. The first three variables to enter the analysis as most discriminant among quartiles of prevention effectiveness are, as stated earlier, features of fire service organization—building inspection program comprehensiveness, education and experience requirements exacted of those who would hold inspection positions, and the number of full-time paid fire department personnel. The changes in Wilks's Lambda in Table 6 indicate the relative contribution of each of the discriminant variables to the discrimination among the four groups. When building inspection program comprehensiveness is used as the initial discriminating variable, Wilks's Lambda is reduced from 1.0 to 0.89. By definition, the size of the reduction in Wilks's Lambda decreases with each new step up to but not including an additional step which is not statistically significant at a specified level. In total, Wilks's Lambda is reduced by 53% (from 1.0 to 0.47) by all 10 variables; the organizational variables contribute 33% and the environmental constraints 20% of the reduction. The statistics in Table 6 do confirm the interpretation proposed earlier, i.e., that the low ratio of alarms to population of the most effective cities probably reflects the community environment ("suburban," middle- and upper-middle-class towns) rather than some kind of fire prevention program. In fact, the rather weak prevention organization characteristic of the most effective towns also reflects the demand for prevention, i.e., rather low.

As mentioned earlier, three successive orthogonal functions were derived. The discriminating power of each can be assessed by reference to its canonical correlation coefficient: 0.51, 0.46, and 0.42. All three coefficients are statistically significant at the level of .007 or better. Since the remaining variation that the second, then the third, functions can possibly explain is reduced, the similarity of size of all three coefficients is unusual. The relative percentage of contributions by each is not that different. Yet by definition, of

TABLE 7
PERCENTAGE OF CORRECT PREDICTIONS OF LEVEL OF PREVENTION EFFECTIVENESS

	No. of Cases	Most Effective	Moderately Effective	Less Effective	Least Effective
Most effective	43	60.5	7.0	2.3	30.2
Moderately effective	44	18.2	61.4	11.4	9.1
Less effective	43	11.6	20.9	51.2	16.3
Least effective	44	11.6	32.6	14.0	41.9
Total	173				

Percentage of correctly predicted cases: 53.8%.

course, the first function always has the largest coefficient. In this case it is 0.51, which suggests that approximately 26% of the variation in level of prevention effectiveness is accounted for by the first linear combination (determined by squaring the coefficient).

Table 7 presents the percentage of correct predictions or classifications. Examination of the cells in the diagonal from upper left (most effective) to lower right (least effective) reveals that over 60% of the cases that were actually in the upper two quartiles of prevention effectiveness were classified in their proper quartile. Slightly over half of the less effective cities were predicted to be so, and only 41.9% of the least effective towns were classified as least effective by the three linear equations. Clearly, the discriminant ability of the 10 variables is strongest for the two more effective quartiles. With respect to the three combinations of these variables, each of these two categories tends to be homogeneous and unlike the others. Overall, the level of effectiveness of 53.8% of the cases was correctly predicted.

FIRE SUPPRESSION EFFECTIVENESS

A total of 11 environmental constraints and nine organizational variables[12] was chosen to discriminate among four levels of fire suppression effectiveness, defined as property loss due to fire per 1,000 population. However, of these 20 variables, only 10 met the stipulations of statistical significance. These variables and their means are listed in Table 8 in the order of their discriminating power.

The category of most effective suppression is characterized on the average by a flexible or changing manning policy (or an improbability of constant manning), fewer thunderstorms, higher probability of offering incentives for fire fighters to pursue further education, a markedly better water supply rating, less comprehensive fire service planning, smaller population size, full-time paid fire fighters, a lower maximum response time, less housing deterioration, and less land area. Table 8 also provides measures of the relative strength of each variable in distinguishing cities grouped into quartiles of suppression effectiveness.

The dummy variable constant manning (whether a fire department assigns a standard number of fire fighters per apparatus on each shift or maintains variable manning) is clearly the strongest discriminatory variable, reducing Wilks's Lambda from 1.00 to 0.85. As the means of this variable indicate, the fire departments of cities which are in the "most effective" quartile in terms of their suppression capabilities apparently assign more personnel to apparatus in places and at times of greater probable hazard and fewer personnel for lesser probable hazard. Constant manning appears both expensive (see Table 4) and ineffective (see Table 8). This pattern could hardly reflect just department and city size, since number of full-time employees and population both appear in the analysis. Even when population size is controlled for (it appears in the denominator of the dependent variable and as a discriminant variable), constant manning appears to be associated with less rather than more effectiveness in fire suppression. Thunderstorms is next in importance and may represent, as the averages of the groups seem to suggest, the difficulty of fighting fires during these storms rather than the difficulty of fighting fires caused by the lightning which often accompanies such storms.

Education incentives, as an indicator of fire department professionalism, enters in step number three and confirms the general belief that more "professional" departments tend to be more effective in reducing property damage due to fire. Of apparently equal importance is water supply; the better departments have a more adequate supply. However, having a more elaborate 5- to 10-year fire service plan in no way guarantees the achievement of a low ratio of fire loss per 1,000 population; the most effective departments are not likely to have a comprehensive formal plan. Each of the last five discriminant variables reduces Wilks's Lambda significantly, but only by two or three points. Two of these, population size and number of full-time personnel, have already been discussed. The remaining three in Table 8 are consistent with expectations.

The canonical correlation coefficient of only the first function is

TABLE 8
MEANS OF SIGNIFICANT VARIABLES IN STEPWISE DISCRIMINANT ANALYSIS OF SUPPRESSION EFFECTIVENESS

	Most Effective	Moderately Effective	Less Effective	Least Effective	All Cities	Wilks's Lambda[a]
1. Constant manning	0.33	0.78	0.86	0.82	0.80	0.85
2. Thunderstorms	21.7	18.2	26.0	30.2	24.6	0.81
3. Further education incentive for fire fighters	0.33	0.05	0.07	0.05	0.07	0.77
4. Water supply	44.8	210.6	218.3	157.2	189.3	0.73
5. Fire service planning	0.23	0.52	0.95	0.68	0.70	0.70
6. Population size (in 1,000s)	81.34	100.94	103.90	139.38	112.77	0.68
7. Number of full-time paid fire fighters	10.33	141.93	153.04	255.45	174.29	0.65
8. Maximum response time	4.67	7.00	7.14	7.47	7.08	0.62
9. Housing deterioration	0.02	0.49	0.73	1.29	0.80	0.60
10. Land area (square miles)	56.17	40.94	63.77	61.11	55.47	0.57

a. All steps are significant at the .003 level or better.

TABLE 9
PERCENTAGE OF CORRECT PREDICTIONS OF LEVEL OF SUPPRESSION EFFECTIVENESS

	No. of Cases	Most Effective	Moderately Effective	Less Effective	Least Effective
Most effective	49	42.9	4.1	4.1	49.0
Moderately effective	50	14.0	42.0	28.0	16.0
Less effective	49	10.2	26.5	42.9	20.4
Least effective	49	12.2	12.2	22.4	53.1
Total	197				

Percentage of correctly predicted cases: 45.2%

statistically significant at the level of .05 or better; it is 0.55 and therefore accounts for about 30% of the variation among levels of suppression effectiveness, approximately two-thirds of all the variation explained by all three functions. This highly significant canonical correlation coefficient ($p < .001$) suggests that level of suppression effectiveness is in part a function of the discriminant variables included in the equation. This conclusion is sharpened by use of the discriminant function to predict the suppression effectiveness level of each municipality. The percentage of correct predictions appears in Table 9.

In contrast to the clear success in identifying the variables, in terms of which each category tends to be relatively homogeneous and dissimilar from the others, the effort to predict the suppression effectiveness level of each city meets with only moderate overall success; 45.2% of the cases are correctly predicted. Of the most effective cities, 42.9% were correctly predicted, but almost half were predicted to be in the least effective category. Only 42% of the moderately effective and almost 43% of the less effective were correctly classified. However, well over half of the least effective cities were classified as least effective. It seems clear that while the least effective cities seem fairly distinctive from all others, the most effective cities differ significantly only from the moderately and less effective groups, but not from the least effective.

FIRE SERVICE PRODUCTIVITY

Earlier it was suggested that the minimization of either loss or expenditure alone and without regard to the value of the other is not enough. Fire service productivity is defined here as the ratio of total

TABLE 10
MEANS OF SIGNIFICANT VARIABLES IN STEPWISE DISCRIMINANT ANALYSIS OF FIRE SERVICE PRODUCTIVITY

	Most Productive	Moderately Productive	Less Productive	Least Productive	All Cities	Wilks's Lambda[a]
1. Emergency response versatility	–0.57	–0.16	0.39	0.45	0.23	0.87
2. Dependent youth	12404.5	10004.8	15284.8	14073.4	13393.2	0.67
3. Maximum response time	3.75	7.50	6.43	8.00	7.17	0.61
4. Full-time paid fire fighters	15.5	102.9	184.2	234.6	175.7	0.55
5. Status of fire chief	2.25	2.60	2.64	2.66	2.62	0.52
6. Number of municipal inspection agencies	1.00	0.92	1.00	1.00	0.98	0.46
7. Social class	29.0%	31.8%	27.1%	23.1%	26.9%	0.45
8. Inspection program comprehensiveness	3.25	6.77	9.43	6.92	7.59	0.42
9. Constant manning	0.50	0.81	0.91	0.82	0.83	0.39
10. Industrialism (in 1000s)	39.9	69.1	129.7	113.2	104.9	0.38
11. Institutionalized population	658.7	848.4	1708.4	1259.0	1284.7	0.36
12. Education incentive for fire fighters	0.25	0.15	0.06	0.05	0.09	0.34
13. Housing deterioration	0.03	0.24	1.28	1.19	0.94	0.33
14. Mutual aid responses	1.50	10.35	44.57	15.18	23.42	0.31
15. Emergency medical/ rescue services	0.37	–0.04	0.16	0.11	0.10	0.30
16. Density	3367.4	3852.9	3954.8	4565.7	4131.6	0.29
17. Inspection staff size	2.5	9.0	16.7	11.0	12.1	0.28
18. Manager form of government	0.75	0.65	0.66	0.55	0.62	0.27

a. All steps are significant at the .004 level or better.

cost to population. What municipalities probably (should) strive for is to attain and maintain a position in the first quartile of productivity, i.e., comparatively low total cost of fire. Table 10 presents the means of 18 variables (of the 34 tried) which made significant contributions to discrimination among four quartiles of fire service productivity.[13]

Eleven of the 18 significantly discriminant variables describe organizational features of the fire service which are manipulable; the other seven variables are environmental constraints. In general, the municipalities that have minimized the total cost of fire have a lesser emergency response versatility, have a lower maximum response time, have fewer full-time employees, are less likely to have a full-time paid chief, have only one municipal agency performing inspections, have a less comprehensive inspection program, are less likely to use constant manning, are more likely to provide incentives to fire fighters to get more education, have fewer mutual aid calls, offer more emergency medical/rescue services, and have a smaller municipal inspection staff. In addition, they have fewer children under seven years of age, have a higher social class, are less industrialized, have a smaller institutionalized population, have less housing deterioration, are less dense, and tend to have a council-manager form of government. An examination of Table 10 indicates the relative importance of each of these variables.

The first five discriminant variables, and especially the first two, clearly make the greatest individual contributions to discrimination among levels of fire service productivity. Four of these five are fire department or suppression organizational characteristics; variables six and eight are prevention organization features. It is evident from these and the numerous other significant variables in the table that level of fire service productivity captures and varies in terms of both prevention and suppression activities. In general, when municipalities are grouped in quartiles according to their total costs of fire, each quartile tends to be homogeneous and unlike the other three. The extent to which the 18 variables identify the group homogeneity and intergroup dissimilarity is revealed in the size of the first two statistically significant canonical correlation coefficients: 0.73 and 0.53. Although the third discriminant function is not statistically significant, the first two are significant and strong. The first linear function alone accounts for over 53% of the variation among levels of fire service productivity; the second explains about 28%. Certainly these two linear combinations are closely associated with variations

TABLE 11
PERCENTAGE OF CORRECT PREDICTIONS OF LEVEL OF FIRE SERVICE PRODUCTIVITY

	No. of Cases	Most Productive	Moderately Productive	Less Productive	Least Productive
Most productive	53	50.9	41.5	0.0	7.5
Moderately productive	53	5.7	58.5	22.6	13.2
Less productive	53	5.7	20.8	56.6	17.0
Least productive	53	3.8	13.2	20.8	62.3
Total	212				

Percentage of correctly predicted cases: 57.1%

in level of productivity. The final test is to determine how well the functions can correctly classify each municipality. Table 11 indicates substantial success.

Again focusing on the diagonal from upper left to lower right, it can be seen in Table 11 that the discriminant functions are least effective in correctly predicting the most productive cities and most effective for those cities whose total cost of fire is highest. Even so, fully 50.9% of the most productive municipalities are correctly classified. If one adds the misclassifications in the adjacent category, the total of correct and near correct is well over 90%. Well over 60% of the least productive cities are correctly predicted. In fact, the overall success of prediction of all cities is 57.1%. With respect to levels of productivity, each quartile is fairly homogeneous and unlike the other three.

VI. SUMMARY AND CONCLUSIONS

This research is an exploratory descriptive exercise, and the findings should be regarded as tentative; no causal inferences can be drawn. Yet the results, however tentative, do suggest hypotheses, that are theoretically interesting and relevant to policy-making, about the relationships among environment, organization, and the delivery of fire protection services.

Performance of the fire service delivery system was measured with four ratios: prevention effectiveness, suppression effectiveness, expenditures, and productivity. In all four cases, use of discriminant analysis did indicate significant environmental and fire service organizational differences among levels of performance. The three general hypotheses stated earlier are clearly supported by the data. Although the supply of fire service output does reflect urban environmental demand, the service delivery performance measures also seem to vary in terms of several service organizational features. How the service is delivered, by what kinds of personnel, and in what kinds of organization all seem to make a difference in the way demand and supply are articulated and in the impact of fire service policy on urban environment (demand) characteristics.

A reexamination of Tables 4, 6, 8, and 10 will provide a summary of which particular variables significantly discriminate among the four performance measures. Perhaps the best general indicator of the significant differences among performance levels is the accuracy of predictions of city performance level by the linear discriminant functions. The best prediction was for levels of productivity, 57.3%. It was not surprising that fire service productivity is most closely associated with the relevant environmental and organizational variables, because it was earlier hypothesized that there is a tradeoff between expenditures and losses. Add to this the fact that the productivity measure both incorporates elements of prevention, suppression, and expenditure and assumes that the total cost of fire must be minimized. It is this total cost of fire or fire service productivity that apparently has the most sensitive linkage with characteristics of both the environment that it serves and the organization that delivers it. It is this interactive organization-environment linkage which captures both the supply-demand relationship and the intended and desirable consequences of public policy delivered to the environment in response to environmental needs and conditions. Prediction was least successful for suppression effectiveness (45.2% correct); the next worst prediction was for prevention effectiveness (53.8% correct). However, expenditures per capita were correctly predicted for 57.3% of the cities. In the cases of prevention and suppression effectiveness the best performers exhibit quite low magnitudes of effort and organizational capability —the opposite of what might be expected. However, several of the environmental characteristics of cities in the high performance categories of prevention and suppression effectiveness clearly indicate that they have less fire hazard and therefore lower incidence and loss. The fire service organizational features of these cities reflect the environmental demand (low) just as their prevention and suppression effectiveness levels do.

With respect to the policy implications of this analysis, if decision makers in cities want to maximize productivity by minimizing the total cost of fire, then several very tentative suggestions emerge from the productivity analysis. The problem of a city in the least productive quartile is illustrative. This city has extremely poor and deteriorated housing, is extremely densely populated, is heavily industrialized, has a relatively large population of children under seven years of age, has the lowest social class of any quartile, and has a large institutionalized (and therefore particularly vulnerable) population. In short, its environment is very hazardous and demanding; little if any short-run changes can be expected in the environment. What can the city do to increase its productivity? Although our analysis is not designed specifically to answer this question, it does suggest some related tentative hypotheses that should be tested in later research. Decision makers in the low productivity city may want to consider the following changes in order to try to improve: (1) make the fire department more professional by such practices as providing incentives to fire fighters to get more education (e.g., pay for the time in school), (2) decrease maximum response time, (3) assign the number of personnel to apparatus on a variable rather than constant ratio (e.g., high ratio during times and in areas of high demand, otherwise a concomitantly lower ratio), and (4) reduce the number of mutual aid agreements. None of these changes is easy, but all are possible. Each represents a feature of organization in which the least productive municipalities are most distinct from the more and certainly most productive towns. However, there is absolutely no guarantee that such organizational modifications would, in fact, reduce the total cost of fire per capita in the short run; the environmental demands on fire service delivery systems in the least productive cities may be simply impervious to any attempted organizational solution in the short-to-intermediate term.

It may never be possible to assert confidently that any given organizational changes will actually produce desired changes in levels of fire service effectiveness or productivity. One possible way to begin is to use multiple-partial correlation and regression techniques, which allow examination of the relationships between performance and organizational variables while holding constant the influence of relevant environmental variables on performance measures. This technique would help to determine whether organizational variables remain significantly related to performance after the influence of environmental constraints is controlled for. An additional approach that could be used involves regression of a performance measure on environmental variables to produce predicted performance values for each case. The difference between the actual and predicted performance measures (the residual) is a measure of the extent of environmental "indeterminism," i.e., the behavior of performance that is not constrained by the environment. Presumably, some of this difference between actual and (environmentally) predicted performance is due to the influence of organizational characteristics of the fire service. The residual could be regressed on organizational variables to determine to what extent this is true (as measured by the R^2) and which organizational characteristics are more important (as measured by standardized beta coefficients). Such analyses will shed light on the extent to which organization makes a difference in performance.

However, an alternative to the organizational changes already suggested is implicit in the findings. It is likely that for high demand municipalities more may not be better. The fact that the least productive cities tend to have many fire service organizational characteristics that one would expect to be associated with high productivity is suggestive. Maybe a reallocation of resources and capabilities is required. For example, decision makers in low productivity municipalities could consider allocating a larger proportion of their existing suppression capability to the densely populated neighborhoods with deteriorated housing. Municipal codes could be passed and enforced, mandating the installation of a variety of private prevention and suppression devices (automatic sprinklers, smoke and heat detectors, alarms, and the like) in new or even existing buildings of numerous kinds, including residences. The point is that extreme demand conditions may not require increases in organizational features—such as inspections, number of fire fighters, professionalism, emergency response versatility, and so on—to produce an appropriate supply of fire service productivity delivered. What may be required is an innovative reallocation of resources.

NOTES

1. The research reported in this chapter represents the initial analysis of results from Phase I of a two-phase study. Phase I was designed to collect and analyze aggregate data on a large sample of cases in order to describe the fire service in America and to evaluate patterns of service delivery, particularly with respect to effectiveness and efficiency. Phase II is

designed to examine more intensively a variety of fire service delivery characteristics in a small number of case studies.

2. Phase II of this ongoing research program will evaluate the equity and responsiveness of fire service delivery systems.

3. Other variables used include percent nonwhite, per capita retail sales, property tax revenues as a percentage of total general revenues, percent of city employment in manufacturing, ratio of intergovernmental revenues to total general revenues, median value of owner-occupied or all housing units, recent population increase, and ratio of central city population to urban fringe population.

4. Hitzhusen (1972) used six different expenditure numerators and their various combinations: annual operating costs, and annual operating costs plus each of the following separately—annual charge for capital, for volunteer effort, for water supply, private fire insurance costs estimated from projected premiums, and private fire insurance costs estimated from key rate and property value data. Two different denominators were constructed: population protected, holding constant the value of real property per person, and the unit of property value protected, holding constant the number of people per unit of property value protected.

5. A summer drought, a big warehouse fire, or a natural disaster can radically alter the effectiveness or productivity measure of a city, even when three-year averages of performance measures are used.

6. A discriminant function assumes the following form:

$$D_i = d_{i1} Z_1 + d_{i2} Z_2 + \ldots d_{ip} Z_p$$

where D_i is the score of a case on discriminant function i, the d's are the weighted coefficients, and the Zs are standardized values of the p discriminating variable in the analysis. See Nie et al. (1975: 435).

7. Classification is achieved through the use of a separate classification function for each group. The classification equations are derived from the pooled within-groups covariance matrix and the centroids (the mean discriminant scores for each group on the respective functions) for the discriminating variables. The classification coefficients are multiplied by the relevant raw variable values, summed, and added to a constant. An equation for one group would take the following form:

$$C_{i0} = c_{i1} V_1 + c_{i2} V_2 + \ldots c_{ip} V_p + c_{i0}$$

where c_{i0} is the classification score for group i, the c_{ij}'s are the classification coefficients, c_{i0} is the constant, and the Vs are the raw variable values. See Nie et al. (1975: 445-446).

8. In the stepwise analysis, the test for entry was the statistical significance of the amount of discrimination added by the variables over and above the discrimination effected by those variables already entered. A default value of at least 1.0 was stipulated for the F-ratio for all new entries because the number of degrees of freedom, cases, and variables entered (upon which F depends) change with each new entry. The F-ratio of 1.00 maintained significance levels of 0.008 or better in all cases.

9. See Research Triangle Institute (1975: 234-236). In an equation with alarms as the dependent variable, number of building inspections has an elasticity of +0.20, i.e., an increase of 100% in the number of inspections would apparently produce an *increase* of 20% in the number of alarms, rather than a decrease as one would hope. Since alarms and loss are positively related, prevention spending and suppression are also positive in association.

10. The 13 environmental variables included are: land area, housing deterioration, crowding, precipitation, windy conditions, thunderstorms, very cold climate, density, property value, city manager government, social class, institutionalized population, and number of fire alarms. Organizational variables suspected of influencing the behavior of expenditures across municipalities are: status of fire chief, unionism, number of full-time

paid employees, magnitude of emergency rescue and medical services offered, constant manning, number of mutual aid calls responded to, and number of alarms answered in contract service areas.

11. The environmental variables included are land area, density, housing deterioration, crowding, very cold climate, thunderstorms, property value, industrialism, social class, and manager form of government. The organizational variables included status of fire chief, full-time paid personnel, fire service planning, building inspectors' training, building inspectors' education and experience, building inspection program comprehensiveness, inspection staff size, and fire fighter inspections.

12. The environmental variables are land area, density, total population, housing deterioration, crowding, windy conditions, thunderstorms, property value, industrialism, social class, and water supply. Status of fire chief, maximum response time, unionism, number of full-time paid fire fighters, fire service planning, emergency response versatility, further education incentive for fire fighters, constant manning, and administrative size are the organization variables.

13. Variables which could have entered the discriminant analysis of productivity but which did not because they failed to meet the test of statistical significance were: experience and education of building inspectors, training of building inspectors, hours of training per month for fire fighters, unionism, administrative size, fire service planning, water supply, crowding, precipitation, very cold days, thunderstorms, windy conditions, land area, property value, contract alarms answered, and number of fire department fire inspections.

REFERENCES

AHLBRANDT, R. S., Jr. (1973) "Efficiency in the provision of fire services." Public Choice 16 (Fall): 1-15.

BAHL, R. W. (1969) Metropolitan City Expenditures: A Comparative Analysis. Lexington: University of Kentucky Press.

--- and R. J. SANDERS (1965) "Determinants of changes in state and local government expenditures." National Tax Journal 17: 50-57.

BRAZER, H. E. (1959) City Expenditures in the United States. Occasional Paper 66. New York: National Bureau of Economic Research.

CHAIKEN, J. M. and J. E. ROLPH (1970) "Predicting the demand for fire service." New York: Rand Institute.

CZAMANSKI, D. A. (1975) The Cost of Preventive Services. Lexington, Mass.: D.C. Heath.

FISHER, G. W. (1964) "Interstate variation in state and local government expenditure." National Tax Journal 17 (March): 57-74.

HIRSCH, W. Z. (1970) The Economics of State and Local Government. New York: McGraw-Hill.

--- (1959) "Expenditure implications of metropolitan growth and consolidation." Review of Economics and Statistics 41: 232-241.

HITZHUSEN, F. J. (1972) "Public-private fire protection cost trade-offs in Texas and New York: a benefit-cost analysis." Ph.D. dissertation. Ithaca, N.Y.: Cornell University. (unpublished)

International City Management Association [ICMA] (1974) Measuring the Effectiveness of Basic Municipal Services. Washington, D.C.: ICMA.

MASOTTI, L. H. and D. R. BOWEN (1965) "Communities and budgets: the sociology of municipal expenditures." Urban Affairs Quarterly 1 (December): 39-58.

NIE, N. H., D. H. BENT, and C. H. HULL (1975) Statistical Package for the Social Sciences. New York: McGraw-Hill.

PIDOT, G. B. (1969) "A principal components analysis of the determinants of local government fiscal patterns." Review of Economics and Statistics 51: 176-188.

PULSIPHER, A. G. and J. L. WEATHERLY, Jr. (1968) "Malapportionment, party competition, and the functional distribution of government expenditures." American Political Science Review 62 (December): 1207-1220.

Research Triangle Institute [RTI] (1975) Evaluating the Organization of Service Delivery Systems: Technical Report of Phase I. Research Triangle Park, N.C.: Research Triangle Institute.

SACKS, S. and R. HARRIS (1964) "The determinants of state and local government expenditures and inter-governmental flows of funds." National Tax Journal 16: 75-85.

SCHMANDT, H. J. and R. STEPHENS (1960) "Measuring municipal output." National Tax Journal 13: 369-375.

SHAENMAN, P. and J. SCHWARTZ (1974) Measuring Fire Protection Productivity in Local Government. Boston: National Fire Protection Association.

SYRON, R. F. (1972) "An analysis of the collapse of the normal market for fire insurance in substandard urban case areas." Research Report No. 49. Boston: Federal Reserve Bank of Boston.

VAN DE GEER, J. P. (1971) Introduction to Multivariate Analysis for the Social Sciences. San Francisco: W. H. Freeman.

WILL, R. E. (1965) "Scalar economics and urban services requirements." Yale Economic Essays 5 (Spring): 3-61.

10

Police Patrol in Metropolitan Areas—Implications for Restructuring the Police

ROGER B. PARKS

□ FOR MANY YEARS, critics of American policing have argued that its fragmented nature leads to poor performance. They have argued that massive restructuring, principally the consolidation of the many smaller local police agencies found in metropolitan areas into one or a few large units, is necessary to improve police performance in metropolitan areas.[1] In spite of this consensus among the critics, most local communities have resisted consolidation efforts.[2] The critics have seen this as an example of the irrationality of residents of those communities. However, recent research has provided evidence that the levels of police service provided to citizens in areas served by small to medium-sized police agencies may be higher than the levels of service provided to residents of similar areas served by larger police agencies.[3] If this is the case, citizens who resist attempts to consolidate their police forces may not be irrational, but rather may simply wish to preserve the levels of service they currently receive.

AUTHOR'S NOTE: *The author gratefully acknowledges the support of the National Science Foundation (Research Applied to National Needs) in the form of Grant GI 43949. The findings and opinions expressed are solely those of the author and do not necessarily reflect those of the Foundation. The author also acknowledges the very helpful critiques and comments provided by Elinor Ostrom, John P. McIver, and Jnana Hodson.*

What *are* the relationships between the multiplicity and/or the fragmentation of a metropolitan police industry and how well police agencies perform? A commonly recommended reform, consolidation of all or some of the police agencies serving a metropolitan area, would result in the dominance of the area's police industry by a single producer.[4] What is the relationship between the degree of police industry domination and police performance? Specialized production of auxiliary services is frequently suggested as a means of capturing possible economies of scale in the production of such services.[5] What relationships exist between the specialization of auxiliary service production in metropolitan areas and the performance of police agencies in those areas?

To answer such questions, several types of data are required. First, comparative measures of metropolitan police industry *structure* are needed to determine which metropolitan areas have higher levels of multiplicity, which are more fragmented, and which are more dominated by a single producer. Relative measurement of the specialization of auxiliary service production is also necessary. Second, one needs measures of the service conditions affecting the demand for police services in metropolitan areas in order to compare areas where such conditions are relatively similar. Most importantly, comparative measures of the performance of police agencies are needed for metropolitan areas exhibiting similarities and differences in police industry structure and service conditions.

Collecting and organizing such information is not a simple task. Service condition data for metropolitan areas and units within them are available from Census Bureau and other agencies' publications. But questions of data currency and the levels of data aggregation still present problems. Also, factors affecting the demand for police service are not clearly identified. Considerably less data are available for characterizing the structure of the police industries that serve various metropolitan areas. There are virtually no data upon which to base a comparison of the performance of police agencies within single metropolitan areas or across metropolitan areas having different industry structures or service conditions.[6]

In this chapter, data from a recent comparative study of the organization of police service delivery in small and medium-sized metropolitan areas are marshalled to provide a preliminary exploration of some of these important relationships.[7] A number of comparative measures of metropolitan structure are presented and employed in the analyses. The research project collecting these data was, by design, a *descriptive* exploration of the current state of policing in metropolitan areas. No attempt was made to collect comparable performance data. There is little agreement among scholars or the police on any measures of police performance that might be employed for such a wide-ranging comparison. Although the lack of performance data severely limits the depth of the analyses, this chapter will present some limited explorations using indicators of police activity levels as proxies for performance.

The analyses are based on data obtained in 80 metropolitan areas across the United States.[8] These 80 areas are randomly selected from the 200 metropolitan areas defined by the Census Bureau in 1970 having a population of less than 1.5 million persons and boundaries that did not cross state lines. The sample was stratified for each of the 10 regions used by the U.S. Department of Justice and other federal agencies.

The 80 metropolitan areas range in size from Meriden/Connecticut (population 55,959) to Paterson-Clifton-Passaic/New Jersey (population 1,357,930). They are located in 31 states. Population densities in the areas range from 31 to more than 3,000 persons per square mile. Over 23 million Americans resided in these 80 areas in 1970. More than 50,000 full-time sworn police officers are engaged in producing police services in the areas. These officers are employed by over 1,400 police agencies. While the very largest metropolitan areas are excluded from consideration, the patterns found in these 80 metropolitan areas are perhaps typical of the range of variation in police service production in areas of the United States other than those few megalopoli.

POLICE INDUSTRY STRUCTURE IN METROPOLITAN AREAS

The 80 metropolitan areas vary quite dramatically in the ways that police are organized. In Meriden/Connecticut, only one agency produces patrol service. In Paterson-Clifton-Passaic/New Jersey, more than 90 agencies do so. The median metropolitan area has 13 patrol producers. But simple counts of numbers of patrol producers do not facilitate comparisons of different metropolitan areas. In order to accomplish the latter, a more complex means for characterizing the structure of the police industry serving each metropolitan area is required.

In the research project collecting the data reported in this chapter, a methodology for such characterization was developed. This methodology requires that each service produced by or for police agencies be considered separately. For a particular service, say the investigation of residential burglaries, the agencies which produce the service and those organized groups which consume the service are arrayed respectively as the rows and columns of a matrix. At the intersection of a row representing a given producer with the column representing a consuming group, any of a series of entries can be made to specify the type of relationship between producer and consumer. Manipulation of matrices generated in this fashion allows the computation of a variety of measures of the structure of organizational arrangements in a metropolitan area for the production of the service in question.

Details of the methodology are presented in several publications that are currently available or that soon will be available.[9] Rather than attempting to spell out the intricacies here, a few of the measures that can be derived through use of the methodology will be presented and used in the subsequent analyses.

As noted, each service produced by or for police agencies is treated separately. A distinction is made between direct police services—produced by police agencies directly for a citizen clientele —and auxiliary services—produced by a police or other agency for consumption by a police agency. In the case of auxiliary services, police agencies that produced direct services become the consuming units or columns in the matrices. Direct services studied were criminal investigation, further separated into investigation of homicides and investigation of residential burglaries; general area patrol; and traffic control, also separated into specific traffic patrol and traffic accident investigation services. Auxiliary services were radio communications, entry-level training, adult pre-trial detention, and chemical analysis. Matrices were prepared and structural measures computed for each of these services. The analyses presented here are focused on the delivery of general area patrol service. Measures of the other direct services will not be included. But a measure of specialization of production for each of three auxiliary services will be included since specialized production of these services is often recommended as a means to improve direct service production.

Three measures of the structure of patrol service delivery will be important for analysis. These are *multiplicity, dominance,* and *autonomy.* Multiplicity is computed as the number of rows in the patrol matrix, or, more simply, as the number of producers of general area patrol service in the metropolitan area. For the purpose of the analyses in this chapter, multiplicity will be stated in relative terms, as the number of producers of patrol service per 100,000 inhabitants of the metropolitan area. Dominance is the proportion of the metropolitan population that receives patrol service from the producer that has the largest "serviced population."[10] Autonomy is measured by the proportion of the metropolitan population in organized groups that receives patrol service *exclusively* from agencies that are bureaus of their local governing authorities.[11] More consolidated metropolitan areas have generally lower values of multiplicity and higher values of dominance. Areas where there is duplicate production of patrol service for the same organized groups have higher values of multiplicity and lower values of autonomy. Metropolitan areas that typify the situation most often criticized by advocates of reforming American police service delivery are those that have high values of multiplicity and autonomy, and a low value for dominance.

Specialization in the production of radio communications, entry-level training, and adult pre-trial detention is inversely measured by the relative multiplicity of service production. This is computed for each service as the number of producers of the auxiliary service divided by the number of direct service producers that consume the service. Metropolitan areas where the production of a given auxiliary service is more specialized have lower values of relative multiplicity for that service.

With comparable measures of the structure of police service delivery in metropolitan areas, one can begin to explore the probable consequences of altering these structures. That is, one can compare areas where the structure more closely resembles that recommended to areas where the structure is more like that commonly criticized. Of course, one can only hypothesize that cross-sectional comparisons of this nature are reflective of results which might be obtained if the recommendations were accepted and structures were altered to conform to the reform model.

PATROL ACTIVITY LEVELS IN METROPOLITAN AREAS

Ideally, one would examine the relationships among measures of the structure of a metropolitan police industry and measures of the performance of the police agencies comprising the industry. By doing this in a comparative way, one could isolate those characteristics of metropolitan police organization which are associated with higher levels of police agency performance, and those characteristics associated with lower performance levels. This, in turn, would provide a grounding for recommendations to alter the structure of policing in metropolitan areas toward forms where higher performance levels can be expected on the basis of the evidence.

Unfortunately, comparative police agency performance data are not available across the 80 metropolitan areas. The problems of using the only available cross-jurisdictional statistics—the annual Uniform Crime Reports—for comparative performance measurement have been documented in great detail.[12] They will not be repeated here. Suffice it to say that police chiefs themselves are quite adamant that such comparisons would be inappropriate.

Without performance measures, one cannot explore structure-performance linkages. But available comparative data do allow the exploration of the relationships between structure and police deployment patterns, and between structure and police patrol density. These latter measures are *not* performance indicators, but there is reason to believe that they are positively related to police performance.[13]

The specific indicator for police deployment patterns is the percentage of total police agency sworn personnel in a metropolitan area actually deployed for street duty on two separate shifts. The specific indicator for police patrol density is the number of sworn personnel on the street per 1,000 metropolitan inhabitants. Although these are not measures of police performance, they are closer, in a process sense, to police output than is the commonly used measure, total sworn officers per 1,000 inhabitants. The National Commission on Productivity recently emphasized the importance of deploying higher proportions of officers for street duty, arguing that this would increase police productivity.[14] To the extent that higher levels of on-street deployment result in increased availability of officers to respond to requests for service, the connection is plausible. But it is important to bear in mind that only limited evidence has linked higher levels of on-street deployment, higher densities of officers on the street, and higher levels of police performance.

Patrol deployment, measured by the percentage of full-time sworn officers actually on the street on the day shift (10 AM) and on the evening shift (10 PM), varied considerably across the 80 metropolitan areas. The average percentage of sworn officers on the street during the day shift was 11%; for the evening shift, 15%, on the average, were on the street. But patrol deployment on the day shift ranged from a low of 5% in one area to more than 18% in another. On the evening shift the range was from 5% to over 25% of total sworn officers actually on the street.

Patrol density, measured by the number of sworn officers on the street per 1,000 inhabitants, also exhibited wide variation. The average density on the day shift was 0.23, approximately one sworn officer on the street for every 4,000 inhabitants. One metropolitan area had less than one officer on the street for every 10,000 inhabitants; two others had at least one officer on the street for every 2,500 inhabitants. On the evening shift, the average patrol density was 0.29, about one officer for every 3,500 inhabitants. The least densely patrolled metropolitan area had one officer on the street for every 8,500 inhabitants.

Two points are important to remember in respect to these metropolitan area measures. First, metropolitan areas are typically composed of a number of individual police agency jurisdictions, with wide variation occurring within metropolitan areas as well as among different metropolitan areas. Comparison of different metropolitan areas allows one to examine the relationships of metropolitan structure, overall metropolitan area deployment, and patrol density. But, it does not reveal the deployment or patrol densities achieved by individual agencies within each metropolitan area. The second point, as noted earlier, is that patrol deployment and density levels are *not* performance indicators. At best, they measure a component of an agency's or a metropolitan police industry's capacity to perform certain services.

METROPOLITAN POLICE INDUSTRY STRUCTURE AND PATROL ACTIVITY LEVELS

Multiple regression analysis has been used to analyze the relationships among measures of metropolitan police industry structure, metropolitan police patrol deployment, and metropolitan police patrol density. Multiple regression analysis is a mathematical technique for fitting a series of observations: a linear combination of two or more variables (the "independent" variables) are related to a third variable (the "dependent" variable). Examination of the function fitted to the data points using this technique can often be used to describe the relationship between any one of the independent variables and the dependent variable while, in a sense, adjusting for the simultaneous relationships between the dependent variable and all of the other independent variables.[15]

In addition to the structural measures for patrol and auxiliary services, the metropolitan population, land area, and number of full-time sworn officers per 1,000 inhabitants are included as independent variables. The latter variable might be considered an indicator of the relative demand for police service in a metropolitan area, after controlling for population and land area requirements. Given two areas with equivalent populations and land areas, the area having more difficult service demands would be expected to have more personnel employed in the production of police services.

Table 1 presents the bivariate correlations among each of the independent variables, and between each of these and each of the dependent variables. No serious multicollinearity problems appeared among the independent variables, at least at the simple bivariate level.[16]

The regression analyses were performed using a stepwise regression procedure.[17] Metropolitan population, land area, and number of full-time sworn officers per 1,000 were entered into the analyses first, then the three measures of patrol structure were entered, and, finally, the measures of other services were allowed to enter or not, depending on their additional contribution to explaining metropolitan area variation. This ordering seems intuitively appealing in the sense of adjusting for metropolitan characteristics that set the stage for patrol service delivery, then for the structure of organization of that delivery, and lastly for additional variables that might affect delivery at the margin, given the other influences.

Table 2 summarizes the regression results for patrol deployment. For each of the variables in the analyses (except metropolitan population and land area) three coefficients are presented. The first of these, b, represents the slope of a straight line relating the independent variable to the dependent variable after adjustment for the influences of the other variables. The second coefficient, s.e., represents the extent of variability or uncertainty in the slope. The third, beta, is a measure of the relative influence of each independent variable in explaining the variation in the particular dependent variable under consideration.[18]

The slope coefficients have a relatively natural interpretation. The first coefficient, expressing the relationship between the number of

TABLE 1
BIVARIATE CORRELATIONS AMONG VARIABLES

	Population	Land Area	Full-Time Sworn per 1,000	Patrol Structure: Multiplicity	Patrol Structure: Dominance	Patrol Structure: Autonomy	Multiplicity For: Training	Multiplicity For: Detention	Multiplicity For: Radio Communications
Population	1.00	0.35	0.13	-0.24	-0.40	-0.28	-0.32	-0.20	0.00
Land Area		1.00	0.18	-0.06	-0.04	-0.16	-0.07	0.11	0.01
Full-Time Sworn Per 1,000			1.00	-0.35	-0.02	-0.16	0.21	0.18	0.32
Patrol Structure									
Multiplicity				1.00	-0.28	-0.08	-0.39	-0.41	-0.55
Dominance					1.00	0.05	0.37	0.30	0.18
Autonomy						1.00	0.23	0.18	0.03
Structure of Other Services - Multiplicity:									
Training							1.00	0.40	0.19
Detention								1.00	0.38
Radio Communications									1.00
Patrol Force									
10:00 AM									
Percent of Sworn on Street	-0.08	-0.11		0.43	-0.31	0.20	-0.33	-0.12	-0.20
Officers on Street per 1,000	0.08	0.07		-0.01	-0.26	0.07	-0.09	0.07	0.14
10:00 PM									
Percent of Sworn on Street	-0.17	-0.09		0.57	-0.23	0.21	-0.27	-0.20	-0.45
Officers on Street per 1,000	0.00	0.12		0.14	-0.21	0.11	-0.06	0.02	-0.04

full-time sworn officers per 1,000 inhabitants and the percentage of full-time officers on the street at 10 AM, indicates that a metropolitan area having one more full-time officer per 1,000 residents than another metropolitan area also has approximately four-tenths of 1% (–0.42%) *fewer* of its full-time sworn officers on the street at 10 AM, after adjusting for the influence of all other variables in the equation. The second coefficient, linking relative patrol multiplicity and patrol deployment on the day shift, indicates that metropolitan

TABLE 2

INFLUENCES ON METROPOLITAN PATROL FORCE DEPLOYMENT—REGRESSION COEFFICIENTS[a]

Variable	Percentage of Full-Time Sworn Officers on the Street - 10 AM			Percentage of Full-Time Sworn Officers on the Street - 10 PM		
	b	s.e.	beta	b	s.e.	beta
Full-Time Sworn Officers per 1,000 Inhabitants	-0.42	0.52	-0.09	-1.64	0.66	-0.25
Patrol Service Structure						
Multiplicity	0.28	0.12	0.32	0.39	0.16	0.32
Dominance	-3.12	1.80	-0.21	-2.74	2.23	-0.13
Autonomy	2.15	0.97	0.23	2.57	1.20	0.20
Other Service Structures						
Training - Relative Multiplicity	-3.04	1.59	-0.23	-1.49	1.97	-0.08
Detention - Relative Multiplicity	3.78	3.17	0.14	---	---	---
Radio Communications - Relative Multiplicity	---	---	---	-3.32	2.31	-0.15
Explained Variation (R^2)		0.34			0.48	

a. Metropolitan population and land area are also in the equation.

areas having one more patrol agency per 1,000 residents than other areas also have approximately one-fourth of 1% (0.28%) *more* of their full-time sworn officers on the street on the day shift.

Dominance measures the proportion of the metropolitan area population served by the producer having the largest clientele and can vary from close to 0.0 in the most atomized area to 1.0 in a completely consolidated area. The coefficient, –3.12, indicates that a metropolitan area where 50% of the population is served by the dominant producer (a proportion of 0.5) will have 0.31% *fewer* of the total sworn officers in the area on the street at 10 AM than will an area where 40% of the population is served by the dominant producer (a proportion of 0.40). The coefficient of autonomy must be similarly scaled to the 0.0 to 1.0 range, and this indicates a higher percentage of sworn officers on the street in metropolitan areas having higher levels of autonomy.

Relative training multiplicity and relative detention multiplicity are also based on ratios that vary from nearly 0.0 to 1.0 or even higher. A smaller percentage of sworn officers is found on the street in areas where the ratio of training producers to police agencies requiring training is high (b = –3.04). A higher percentage of sworn officers on the street is found in areas where the ratio of detention producers to police agencies requiring detention is high (b = 3.78). Stated in terms of specialization, this means that specialization of training service production is positively associated with a higher percentage deployment of sworn officers for street duty, but that detention specialization is negatively associated with higher levels of patrol deployment, after adjustment for the influences of all other variables.

In all cases, metropolitan population and land area have little influence after adjustments for the influences of the other variables are made. In the case of patrol deployment on the day shift, radio communication multiplicity does not contribute to the analysis.

Looking at patrol deployment on the evening shift, the multiplicity for radio communications replaces the multiplicity of detention service in terms of relative contribution to the analysis.[19] Further, while the effect of metropolitan structure does not change markedly, the negative relationship between the total number of officers per 1,000 inhabitants and the percentage of sworn officers on the street becomes much stronger.

The beta coefficients for the variables in the analyses of patrol deployment show that the structural measures of patrol service

TABLE 3
INFLUENCES ON METROPOLITAN PATROL DENSITY—REGRESSION COEFFICIENTS[a]
DENSITY OF PATROL OFFICERS

Variable	Patrol Officers on the Street Per 1,000 Inhabitants - 10 AM			Patrol Officers on the Street Per 1,000 Inhabitants - 10 PM		
	b	s.e.	beta	b	s.e.	beta
Full-Time Sworn Officers Per 1,000 Inhabitants	.099	.011	.77	.096	.014	.67
Patrol Service Structure						
Multiplicity	.004	.003	.19	.009	.003	.36
Dominance	-.058	.039	-.15	-.033	.050	-.08
Autonomy	.058	.021	.24	.075	.027	.27
Other Service Structures						
Training - Relative Multiplicity	-.077	.035	-.22	-.048	.044	-.12
Detention - Relative Multiplicity	.079	.069	.11	.057	.087	.07
Radio Communications - Relative Multiplicity	---	---	---	---	---	---
Explained Variation (R^2)		0.57			0.44	

a. Metropolitan population and land area are also in the equation.

delivery are generally the strongest contributors explaining variations in the percentage of sworn officers deployed for street duty. The structural measures for the other services contribute somewhat less, as does the number of full-time sworn officers per 1,000 residents of the metropolitan areas.

Summarizing the relationships between measures of patrol service structure and the deployment of full-time sworn officers for street duty in the 80 metropolitan areas, the percentage of the total sworn officers actually on the street during either the day or the evening shift is higher in metropolitan areas having higher values of patrol multiplicity, lower values of patrol dominance, and higher values of patrol autonomy. The metropolitan areas with lower patrol multiplicity and higher dominance (that is, those closer in structure to the form recommended by many contemporary police reform experts) generally have lower percentages of sworn officers deployed for street duty on either shift.

How does metropolitan patrol structure relate to the availability of patrol officers for citizens? Table 3 provides evidence on this question, with availability measured by the density of patrol service (the number of patrol officers on the street per 1,000 inhabitants of the metropolitan area). The strongest relationship is between the number of sworn officers in total per 1,000 inhabitants and the number of patrol officers on the street per 1,000. The slope coefficient (b = 0.099) indicates that metropolitan areas having one more sworn officer per 1,000 inhabitants have one-tenth more patrol officers on the street per 1,000 residents. A difference of 10 sworn officers between two metropolitan areas where population and all else are the same would correspond to a difference of one patrol officer on the street on each shift.

The signs of the relationships between measures of patrol and other service structures and the density of patrol service provided in the 80 metropolitan areas are the same as the signs of the respective relationships with patrol deployment. Higher levels of patrol multiplicity and patrol autonomy, and lower levels of patrol dominance, are associated with higher densities of patrol service. Lower relative multiplicity in the production of training, and higher relative multiplicity in the production of adult pre-trial detention, are also associated with higher patrol densities. Here, as with patrol deployment, the measures of patrol structure have relatively stronger influences on the variation in patrol density than do the measures of the structure of other services. Of more importance, the structure of

TABLE 4
CHARACTERISTICS OF LOCAL PATROL AGENCIES IN METROPOLITAN AREAS

Metropolitan Patrol Structure	(N)	Size of Local Patrol Agencies - Number of Full-Time Sworn Officers		Percentage of Local Patrol Agencies Producing Their Own:				
		Median	Inter-Quartile Range	Burglary Investi-gation	Homicide Investi-gation	Entry-Level Training	Adult Deten-tion	Radio Communi-cations
All Local Patrol Agencies	1118	9	3 - 27	82	69	6	12	68
Multiplicity								
Low	233	17	6 - 54	85	76	8	14	71
Medium	423	13	5 - 30	86	74	7	11	75
High	462	5	1 - 23	77	60	4	13	59
Dominance								
Low	640	9	3 - 27	84	73	5	10	63
Medium	244	9	3 - 31	87	70	7	16	77
High	234	7	2 - 21	73	56	8	15	71
Autonomy								
Low	414	13	2 - 21	79	70	6	11	62
Medium	436	10	3 - 28	83	67	6	11	73
High	268	9	4 - 31	84	69	8	16	68

service delivery for patrol and for the auxiliary services has an influence over and above the obvious influence of the number of sworn officers per 1,000 metropolitan inhabitants.

POLICE AGENCIES IN THE METROPOLITAN AREAS

In attempting to understand the relationships of the structure of patrol service delivery to the levels of patrol deployment and density in the metropolitan areas, a focus on the types of agencies found in the areas is useful. In Table 4, some of the characteristics of the local agencies producing patrol service are arrayed for metropolitan areas that vary on the structural measures.[20] Looking at the variation in patrol multiplicity, the median size of the local patrol agencies in a metropolitan area decreases as the multiplicity of patrol service production increases. Median agency size also is lower in areas having higher levels of dominance and of autonomy in patrol service production. Local patrol agencies in areas having high patrol multiplicity are less likely to conduct their own homicide investigations or produce their own radio communications. This is due to reliance on specialized producers for these services by the smaller agencies.

The proportion of local patrol producers that produce their own radio communications is lower in areas exhibiting lower levels of dominance in patrol service production. In such areas, sharing of dispatch centers of contracting is often found among groups of agencies of similar size or between groups of smaller agencies and a county sheriff. In more dominated areas, the proportion of local agencies conducting homicide investigations is somewhat lower. In these areas, specialized investigators are often available from the dominant patrol producer, from a state agency, or from an organized areawide major case squad.

Table 5 shows the relationship between local patrol agency size and propensity to produce additional services beyond general area patrol. Most local patrol agencies of any size also produce traffic control services–traffic patrol and accident investigation. The smaller local patrol producers are less likely to produce their own residential burglary investigations. This is even more true for homicide investigations. Looking at the auxiliary services, a marked relationship between agency size and the production of these services appears.

TABLE 5
PRODUCTION OF ADDITIONAL SERVICES: LOCAL PATROL AGENCIES

Number of Full-Time Sworn Officers	(N)	Percentage of Local Patrol Agencies Producing:							
		Direct Police Services				Auxiliary Police Services			
		Traffic Patrol	Traffic Accident Investigation	Residential Burglary Investigation	Homicide Investigation	Radio Communications	Entry-Level Training	Adult Pre-Trial Detention	Chemical Analysis
All Local Patrol Agencies	1090	93	89	82	69	68	6	12	2
None: Part-Time Only	80	88	75	49	19	27	2	0	0
1 to 4	271	96	91	66	42	39	1	1	0
5 to 10	248	95	91	86	70	67	0	7	0
11 to 20	156	92	90	92	88	88	1	13	0
21 to 50	161	91	89	96	93	94	1	24	1
51 to 150	107	92	91	99	97	96	16	33	5
Over 150	67	94	91	99	99	93	61	34	19

What is interesting about the production of auxiliary services by local patrol agencies is the variation across services. Most local agencies having five or more sworn officers, and nearly all agencies having more than 10 full-time sworn officers, produce their own radio communications, either entirely or for at least one shift. On the other hand, only among the very largest local patrol producers do a majority produce their own entry-level training, and only a small fraction of local patrol producers of *any* size produce their own adult pre-trial detention and chemical analysis services. Chemical analysis, in particular, is the province of highly specialized laboratory producers.

By *not* producing a number of additional services, the smaller local patrol producers are able to achieve a higher level of patrol officer deployment and a higher density of patrol service. Table 6 shows the relationship between size of local patrol agency and patrol deployment. As agency size increases, a lower percentage of sworn officers is assigned to an agency's patrol division, a lower percentage of the patrol division is actually on the street during the day or evening shift, and, thus, a lower percentage of an agency's sworn personnel is deployed for street duty at any given time.

In Table 7, the relationship of agency size and patrol density is shown. Smaller agencies deploy a higher proportion of their officers for duty on the street and thus achieve a higher density of patrol service than do larger agencies. During the evening shift, the median local patrol agency of five to 10 full-time sworn officers has approximately one officer on the street for every 2,300 members of its serviced population, while the median local patrol agency of more than 50 full-time sworn officers has one officer on the street for every 4,300 citizens served. On the day shift, the corresponding figures are one officer per 3,000 served for the median smaller agency and one per 5,400 for the larger.

SPECIALIZATION AND RESTRUCTURING OF METROPOLITAN POLICE SERVICES

It may be useful to conceive of the many small agencies in metropolitan areas as specialists in the production of patrol, traffic control, and, usually, burglary investigations. By specializing in the production of these services to the exclusion of others, these agencies are able to maintain a higher presence in the communities they serve.

TABLE 6
DEPLOYMENT OF SWORN OFFICERS: LOCAL PATROL AGENCIES

Number of Full-Time Sworn Officers	Percentage Assigned to Patrol Division		Percentage of Patrol Division on the Street				Percentage of Sworn Officers on the Street			
			10 AM		10 PM		10 AM		10 PM	
	Median	Inter-Quartile Range	Median	Inter-Quartile Range	Median	Inter-Quartile Range	Median	Inter-Quartile Range	Median	Inter-Quartile Range
All Local Patrol Agencies	82	64 - 100	22	17 - 33	29	20 - 38	17	11 - 25	20	14 - 33
1 to 4	100	100 - 100	50	25 - 50	50	33 - 100	33	25 - 50	50	33 - 100
5 to 10	100	83 - 100	20	17 - 30	29	20 - 38	20	17 - 25	25	20 - 33
11 to 20	75	65 - 87	22	17 - 27	29	22 - 33	16	13 - 20	20	16 - 25
21 to 50	67	59 - 75	21	17 - 25	25	20 - 30	14	11 - 17	16	13 - 20
51 to 150	60	52 - 68	17	13 - 22	23	18 - 29	10	7 - 13	14	11 - 16
Over 150	53	45 - 59	16	13 - 21	21	16 - 28	9	7 - 11	12	8 - 15

TABLE 7
DENSITY OF PATROL SERVICE: LOCAL PATROL AGENCIES

Number of Full-Time Sworn Officers	(N)	Number of Patrol Officers on the Street per 1,000 Resident Serviced Population			
		Day Shift - 10 AM		Evening Shift - 10 PM	
		Median	Inter-Quartile Range	Median	Inter-Quartile Range
All Local Patrol Agencies	969	0.28	0.16 - 0.50	0.37	0.22 - 0.63
1 to 4	246	0.47	0.17 - 0.86	0.57	0.32 - 1.04
5 to 10	242	0.34	0.13 - 0.55	0.44	0.26 - 0.71
11 to 20	156	0.27	0.17 - 0.45	0.34	0.24 - 0.51
21 to 50	156	0.25	0.15 - 0.37	0.29	0.20 - 0.46
51 to 150	106	0.18	0.12 - 0.25	0.23	0.17 - 0.33
Over 150	63	0.19	0.11 - 0.26	0.23	0.16 - 0.29

This does *not* say that citizens served by smaller police agencies do not receive specialized investigation services when necessary, or that the smaller agencies do not have radio communications, entry-level training, detention, and chemical analysis services available to them. Indeed, a rich structure of interorganizational relationships in most metropolitan areas makes these services available to citizens and police agencies throughout the area. A focus such as that taken in this research, that goes beyond the enumeration of individual agencies, reveals these networks where critics have found a lack of capability for service production. It is true that most small police agencies (and many much larger ones) do not produce highly specialized technical services, but in all parts of the country there *are* producers who have such capability and make these services available as needed by other agencies.

National commissions and other experts concerned with restructuring police service delivery have often recommended that more specialization in the production of auxiliary services would be beneficial. Similar recommendations are made for the maintenance of qualified investigators to handle complex criminal cases involving homicide, rape, hard narcotics, organized crime, and the like.[21] At the same time, many of these commissions and experts have recommended the elimination of what we call here "patrol specialists"—the smaller agencies that allocate most of their resources to on-street activities. The data presented in this chapter may be useful in attempting to analyze the implications of adopting the latter recommendations.

The relationships found indicate that a more consolidated structure of police service delivery in metropolitan areas *may* result in lower levels of on-street deployment of sworn police officers, and in correspondingly lower levels of density of patrol service available to citizens. "May" is emphasized because it is dangerous to make causal or dynamic predictions with cross-sectional data such as these. But the metropolitan areas most closely resembling the model of police service delivery proposed by many reformers turn out to have *lower deployment and patrol density levels.*

As emphasized earlier, higher levels of on-street deployment and/or density of patrol service are not measures of higher police performance. But there is reason to believe that they are positively associated with police performance, other things being equal. Thus, restructuring police service delivery along the recommended lines might well lead to lower levels of police performance, at least for the many functions carried out by the patrol force.

NOTES

1. Many critics have argued these points. Some of the more well-known statements include Advisory Commission on Intergovernmental Relations (1974: 24), Committee for Economic Development (1972), Skoler and Hetler (1970), and President's Commission on Law Enforcement and Administration of Justice (1967: 301).

2. For lists of votes on this issue and their results, see Marando (1975) and National Association of Counties (1973).

3. A number of studies conducted by scholars associated with the Workshop in Political Theory and Policy Analysis have reported such findings. See, for example, Ostrom et al. (1973), Ostrom and Parks (1973), Parks (1976), Ostrom (1976), and Rogers and Lipsey (1974).

4. The Committee for Economic Development is perhaps the strongest advocate of such consolidation (1972: 29-38). Nowhere in their recommendations do they provide any supporting evidence. Ostrom (1975) points out the need for such evidence *prior* to major reforms.

5. David L. Norrgard (1969) is an early advocate of specialization. The National Advisory Commission on Criminal Justice Standards and Goals recently reiterated support for this (1973).

6. The studies cited in Note 3 do provide a basis for comparison of the performance of some police agencies in a limited number of metropolitan areas. There has been to date no comprehensive effort to collect comparable police performance data across a large number of metropolitan areas.

7. This chapter is an early report on data collected by the Police Services Study, a joint project of the Workshop in Political Theory and Policy Analysis at Indiana University and the Center for Urban and Regional Studies of the University of North Carolina. Extensive data analysis is continuing on the study. The analyses reported here should be considered preliminary. They are presented at this time to stimulate comments and criticisms which will improve the on-going analytic efforts of the study.

8. The initial phase of the Police Services Study, which has been completed, was intended to produce descriptive data on the current organization of police service delivery in the metropolitan areas studied. Two volumes reporting findings from the study are currently in preparation and will be available in mid-1976 (see Ostrom, Parks, and Whitaker, 1976a, 1976b).

9. An early statement of this methodology can be found in Ostrom, Parks, and Whitaker (1974). A revised version will be published in Ostrom, Parks, and Whitaker (1976b).

10. Serviced population for a given producer may differ significantly from the producer's nominal jurisdiction. For example, a county sheriff usually has jurisdiction throughout a county but in practice may only produce patrol service for residents of the unincorporated portion of the county. The latter population is what is referred to in this chapter as serviced population.

11. Military base commanders and college administrators are considered local governing authorities in this analysis.

12. Some of the better critiques of the use of crime statistics for comparative purposes are Biderman (1966), Ostrom (1971), and Maltz (1972).

13. Ostrom et al. (1973) speculate that a difference in the type of production strategy adopted by police agencies may partially explain the differences in performance levels that they found. The smaller police agencies in that study allocated a much higher proportion of their resources to patrol activities than did the large police agency, and the former achieved

a two to three times higher density of patrol service. The smaller police agencies were found to be providing equal or higher levels of performance across a wide range of indicators.

14. See National Commission on Productivity (1973). A statement by the commission's chairman which makes this point also is Morgan (1975).

15. Draper and Smith (1966) and Wonnacott and Wonnacott (1970) contain readily accessible discussions of multiple regression analysis. The use of multiple regression as a technique for exploring and describing relationships among variables follows the suggestions of Tufte (1969) and Taylor (1974).

16. Van de Geer (1971: 109) points out that multicollinearity may be a multidimensional problem. Analyses beyond simple bivariate correlations are necessary to determine the extent of any such problems. These have not yet been done with the current data.

17. The particular routine used is that of Nie et al. (1975: 345-347). It consists of a forward selection procedure (Draper and Smith, 1966: 169-171) with no attempt to remove variables already in the equation.

18. Since the analyses presented here are intended to be broadly descriptive, variables are included in the equations which might ordinarily be dropped on statistical grounds. One widely used criterion would suggest dropping all variables where the value of b was not at least twice that of s.e. In the current analyses, variables were retained even though in some cases the value of s.e. equalled or exceeded that of b. With the exception of metropolitan population and land area (variables that were retained in all cases), a variable was dropped only if s.e. exceeded b by more than 50%.

19. The negative coefficients for relative multiplicity of training and radio communication production support arguments for specialized production of these services (Norrgard, 1969; National Advisory Commission on Criminal Justice Standards and Goals, 1973). The positive coefficient for relative multiplicity of detention production does not support the similar arguments for detention specialization.

20. Local agencies are municipal police departments, including those of New England towns and of townships in other states, county sheriffs and police departments, and other police agencies organized at a municipal or county level. Specifically excluded are state police or highway patrols, military police agencies, or other federal producers.

21. In a number of the metropolitan areas studied in this project, organized major case squads were available to immediately assist local agencies with investigations of very serious crimes. Technical reports describing a number of these are available from the project Publications Secretary, Workshop in Political Theory and Policy Analysis, Indiana University, Bloomington, IN 47401.

REFERENCES

Advisory Commission on Intergovernmental Relations [ACIR] (1974) American Federalism: Into the Third Century. Its Agenda. Washington, D.C.: U.S. Government Printing Office.

BIDERMAN, A. D. (1966) "Social indicators and goals," pp. 68-153 in R. A. Bauer (ed.) Social Indicators. Cambridge, Mass.: MIT Press.

Committee for Economic Development [CED] (1972) Reducing Crime and Assuring Justice. New York: CED.

DRAPER, N. R. and H. SMITH (1966) Applied Regression Analysis. New York: John Wiley.

MALTZ, M. D. (1972) Evaluation of Crime Control Programs. Washington, D.C.: U.S. Government Printing Office.

MARANDO, V. L. (1975) "The politics of city-county consolidation." National Civic Review 64 (February): 76-81.

MORGAN, J. P. (1975) "Planning and implementing a productivity program," pp. 129-149 in J. L. Wolfe and J. F. Heaphy (eds.) Readings on Productivity in Policing. Washington, D.C.: Police Foundation.

National Advisory Commission on Criminal Justice Standards and Goals (1973) Report on Police. Washington, D.C.: U.S. Government Printing Office.

National Association of Counties (1973) Consolidation: Partial or Total. Washington, D.C.

National Commission on Productivity (1973) Opportunities for Improving Productivity in Police Services. Washington, D.C.

NIE, N. H., C. H. HULL, J. G. JENKINS, K. STEINBRENNER, and D. H. BENT (1975) SPSS: Statistical Package for the Social Sciences. New York: McGraw-Hill.

NORRGARD, D. L. (1969) Regional Law Enforcement. Chicago: Public Administration Service.

OSTROM, E. (1976) "Size and performance in a federal system." Publius 6.

--- (1975) "On righteousness, evidence, and reform: the police story." Urban Affairs Quarterly (June): 464-486.

--- (1971) "Institutional arrangements and the measurement of policy consequences." Urban Affairs Quarterly (June): 447-474.

--- W. H. BAUGH, R. GUARASCI, R. B. PARKS, and G. P. WHITAKER (1973) Community Organization and the Provision of Police Services. Sage Professional Paper in Administrative and Policy Studies 03-001. Beverly Hills, Calif.: Sage.

OSTROM, E. and R. B. PARKS (1973) "Suburban police departments: too many and too small?" pp. 367-402 in L. Masotti and J. Hadden (eds.) The Urbanization of the Suburbs. Beverly Hills, Calif.: Sage.

--- and G. P. WHITAKER (1976a) Policing Metropolitan America. Washington, D.C.: U.S. Government Printing Office.

--- (1976b) Patterns of Metropolitan Policing. Lexington, Mass.: Lexington Books.

--- (1974) "Defining and measuring structural variations in interorganizational arrangements." Publius 4 (Fall): 87-108.

PARKS, R. B. (1976) "Complementary measures of police performance," pp. 185-218 in K. M. Dolbeare (ed.) Public Policy Evaluation. Beverly Hills, Calif.: Sage.

President's Commission on Law Enforcement and Administration of Justice (1967) The Challenge of Crime in a Free Society. Washington, D.C.: U.S. Government Printing Office.

ROGERS, B. D. and C. M. LIPSEY (1974) "Metropolitan reform: citizen evaluations of performances in Nashville-Davidson County, Tennessee." Publius 4 (Fall): 19-34.

SKOLER, D. L. and J. M. HETLER (1970) "Governmental restructuring and criminal administration: the challenge of consolidation." Georgetown Law Review 58: 719-741.

TAYLOR, C. L. (1974) "The uses of statistics in aggregate data analysis," pp. 68-84 in J. F. Herndon and J. L. Bernd (eds.) Mathematical Applications in Political Science VII. Charlottesville: University of Virginia Press.

TUFTE, E. R. (1969) "Improving data analysis in political science." World Politics 21 (July): 641-654.

VAN de GEER, J. P. (1971) Introduction to Multivariate Analysis for the Social Sciences. San Francisco: W. H. Freeman.

WONNACOTT, R. J. and T. H. WONNACOTT (1970) Econometrics. New York: John Wiley.

11

The Influence of Financial, Educational, and Regional Factors On the Distribution of Physicians

PATRICK O'DONOGHUE

INTRODUCTION

□ THIS CHAPTER IS BASED UPON A LARGER STUDY whose purpose is to analyze health care delivery in metropolitan areas.[1] In particular, this chapter is drawn from the first phase of that study. The central product of this first phase is entitled "Factors Influencing Health Care Resources and Utilization in Metropolitan Areas."[2] Phase I analyzes the following types of health care resources and utilization:

(1) physician geographic and specialty distribution,

(2) concentration of hospital resources,

(3) aggregate hospital utilization patterns, and

(4) concentration and utilization of nursing home beds.

The general subject of this study is those standard metropolitan statistical areas (SMSAs) whose population did not exceed 1.5 million persons in the 1970 census and which did not include jurisdictions for more than one state. There are 200 SMSAs that have

AUTHOR'S NOTE: *The work upon which this article is based was supported by contract NSF C-899 with the National Science Foundation.*

these characteristics. Further, the Minneapolis/St. Paul SMSA was added to the universe of SMSAs for this project.

Using a sample selection technique designed to enhance the analytical soundness of the sample so that the impact of important policy and situational variables could be assessed, 91 SMSAs were selected from the universe of 201 for the Phase I analyses. Further, data incongruities prevented the inclusion of the New England SMSAs in the physician analyses. Consequently, these analyses were conducted using the 80 SMSAs (from the sample of 91) that are not located in New England.[3]

The primary time period for Phase I of this study is the five-year span from 1968 to 1973. Within this block of time, four years received major emphasis–1968, 1970, 1972, and 1973.

The unit of analysis in this project is the SMSA. The SMSA populations were adjusted for the net inflow of hospital patients in order to create population bases that constitute relatively self-contained medical market areas. The SMSA populations were also adjusted for the presence of federal hospitals in the SMSA. Further, adjustments in the SMSA populations across time were conducted using the Census Bureau population estimates.

Two techniques, correlation analysis and profile analysis, were used to assess the degree of bivariate association between two measures. Multivariate relationships were assessed using multiple regression techniques.

The findings discussed in this article are derived from the physician analyses. Those analyses showed that there are three variables which are the primary predictors (from both a substantive and statistical viewpoint) of physician distribution. They are:

(1) physician fee levels,

(2) medical education programs, and

(3) the West region.

This chapter focuses exclusively on the relationships between these three variables and the physician measures.

VARIABLES USED IN THIS ANALYSIS

PHYSICIAN MEASURES

The source of data about physician characteristics is the annual survey of physicians conducted by the American Medical Association (AMA). Drawing upon the data available in this source, the following physician measures are the primary foci of this set of analyses:

(1) the number of non-federal, patient care, office-based physicians per population of the SMSA (in 100,000s);

(2) the number of non-federal, patient care, office-based physicians practicing in surgical specialties per population of the SMSA (in 100,000s); and

(3) the number of non-federal, patient care, office-based physicians practicing in general practice or in the medical specialties per population of the SMSA (in 100,000s).[4]

The rationale for examining the determinants of the concentration of surgical specialists is threefold. First, if the surgical operation rate is strongly dependent on the concentration of surgeons, then increasing concentrations of such physicians may lead to the overprovision of operations, and to higher costs and lower quality of care. Second, there has been serious concern for at least five years that the concentration of specialists is becoming too high and the concentration of primary care physicians (who will be defined shortly) too low. If this is the case, then the result may be insufficient access to care by consumers and unduly high costs.[5] Third, surgical specialists tend to be relatively intensive users of hospital care and other expensive types of medical services. Thus, they play a crucial role in the functioning of the largest segment of the health care industry.

Primary care is usually defined as the provision of medical services at the first point of physician contact with the patient. Measure (3) is directed toward physicians who generally provide this type of care—including both general practitioners and physicians who practice in the medical specialties. The definition of general practice includes only the field of general practice. In contrast, the definition of medical specialties includes a number of different specialties: allergy, cardiovascular diseases, dermatology, gastroenterology, internal medicine, pediatrics, pediatric allergy, pediatric cardiology, and pulmonary diseases.

On substantive grounds it would have been preferable to have included within the primary care physician measure the specialty of obstetrics and gynecology and to have excluded the specialties of pediatric allergy, pediatric cardiology, allergy, gastroenterology, pulmonary diseases, and perhaps cardiovascular diseases. However, the way in which the data are tabulated precludes this approach. Therefore, a relatively good compromise was reached, defining primary care to include general practice and the medical specialties. In this regard it should be noted that measure (3) (primary care) is appropriately dominated by the fields of general practice, internal medicine, and pediatrics.

The rationale for the analysis of the concentration of primary care physicians represents to some extent a mirror image of the rationale for the analysis of the concentration of surgical specialists. First, since primary care physicians are the first point of contact between the consumer and the health care system, the concentration of such physicians is a crucial determinant of the access of consumers to the health care system. Second, if consumers were treated more rapidly and more frequently by primary care physicians, it might be possible to lower the costs of health care by substituting ambulatory care for hospital care. At the same time the quality of health care could be raised by treating illnesses earlier in their course.[6]

The means and standard deviations of the physician variables are shown in the regression equation tables discussed later. They appear as Appendices A, B, C, and D at the end of this chapter.

There are three cohorts of physicians who contribute to changes in the concentration of physicians. The first cohort is composed of physicians entering practice for the first time, most of whom are young. The second cohort is made up of physicians who shift locations and/or specialties in mid-career. The third cohort is composed of physicians who retire or die.

It seems likely that the determinants of these three cohorts are somewhat different. However, since the AMA survey does not permit the specification of these three different cohorts, it is not possible to test this hypothesis directly. Nonetheless, in developing the operational hypotheses for the physician analyses we will consider these three different cohorts and the possible differential impact that independent variables may have on them.

There is another dimension along which physicians can be grouped. This is the dimension as to whether they receive their medical education in the United States or in foreign countries.[7]

TABLE 1
NEW LICENTIATES REPRESENTING ADDITIONS TO THE MEDICAL PROFESSION, BY COUNTRY OF GRADUATION: 1950-1972

Year	Total	U.S. and Canadian	Foreign Medical Graduates		Foreign Medical Graduates as Percentage of Total
			Total	U.S. Born	
1950	6,002	5,694	308	N.A.	5.1
1951	6,273	5,823	450	N.A.	7.2
1952	6,885	6,316	569	N.A.	8.3
1953	7,276	6,591	685	N.A.	9.4
1954	7,917	7,145	772	N.A.	9.8
1955	7,737	6,830	907	N.A.	11.7
1956	7,463	6,611	852	N.A.	11.4
1957	7,455	6,441	1,014	212	13.6
1958	7,809	6,643	1,166	284	14.9
1959	8,269	6,643	1,626	366	19.7
1960	8,030	6,611	1,419	386	17.7
1961	8,023	6,443	1,580	468	19.7
1962	8,005	6,648	1,357	201	17.0
1963	8,283	6,832	1,451	395	17.5
1964	7,911	6,605	1,306	200	16.5
1965	9,147	7,619	1,528	411	16.7
1966	8,851	7,217	1,634	252	18.5
1967	9,424	7,267	2,157	279	22.9
1968	9,766	7,581	2,185	235	22.4
1969	9,978	7,671	2,307	179	23.1
1970	11,032	8,016	3,016	198	27.3
1971	12,257	7,943	4,314	210	35.2
1972	14,476	7,815	6,661	240	46.0

SOURCE: United States Department of Health, Education and Welfare [DHEW] (1974) *Foreign Medical Graduates and Physician Manpower in the United States,* DHEW Publication No. (HRA) 74-30, February 1974.

Table 1 illustrates the sharp increase in the flow of foreign medical graduates (FMGs) to this country. In 1950, 5% of physicians newly licensed and representing additions to the medical profession were foreign medical graduates. During the 1950s this percentage rose slowly, although more rapidly at the end of that decade. Thus, in 1959, 20% of new licentiates were FMGs. This fraction remained relatively constant during the 1960s. For example, in 1968 (the first year of this study), 22% of new licentiates were FMGs. However, the number of FMGs entering the medical profession in this country escalated rapidly during the five years of this study. By 1970, 27% of new licentiates were FMGs, and by 1972, 46% (almost one-half) of new licentiates were FMGs.[8] While FMGs had an impact on physician distribution patterns during the last 25 years, this influence has been much more substantial during the five year period of this study, i.e., from 1968 to 1973.[9]

It seems likely that those factors most heavily influencing the location decisions of foreign medical graduates may differ somewhat from those factors most strongly affecting the location decisions of U.S. medical graduates (USMGs). Consequently, in developing the conceptual framework for these analyses, it is essential to consider the possible differential effects of particular policy and situation variables on FMGs as opposed to USMGs.

INDEPENDENT VARIABLES

There are three independent variables used in these analyses. One is the West region (which corresponds to the West region specified by the Bureau of Census). The mean of this variable is .30 and its standard deviation is .46. This indicates that 30% of the SMSAs in the sample of 80 are located in the West region.

The second independent variable is the concentration of interns and residents, defined as the number of interns and residents per population (in 100,000s). The data source for this variable was the *Directory of Approved Internships and Residencies* published by the American Medical Association. This variable is available for all four study years—1968, 1970, 1972, and 1973. Its mean in 1968 is 11.8, with a standard deviation of 15.2. Its mean in 1973 is 16.4, with a standard deviation of 20.1. Thus, there was an increase of almost 50% in the concentration of interns and residents across this five-year period. Further, as expected, the standard deviation of this variable is large compared to the mean.[10]

A number of physician fee indices were used in the Phase I study. The fee index found most satisfactory on both substantive and statistical grounds was the surgical fee index. The source of data for this index was the Medicare prevailing physician charges data compiled by the Bureau of Health Insurance (SSA) based on data furnished by the Medicare intermediaries. The surgical fee index was constructed by taking a weighted average of prevailing fees for a range of selected surgical procedures, involving different specialty fields. The mean of this variable is $386, with a standard deviation of $60. Medicare prevailing charges data were only obtained for 1973, and, thus, the surgical fee index could only be constructed for 1973.[11]

CONCEPTUAL FRAMEWORK

PHYSICIAN FEE LEVELS

It seems likely that physicians respond to financial incentives (Maloney, 1970). It also seems probable that physicians will weigh increases in income to a progressively lesser extent as income rises. Further, while physicians will probably positively value work, their valuation of work will tend to decrease as the hours worked per week increase.

Therefore, physicians should prefer higher fee levels to lower fee levels, and this preference should be reflected in their geographic and specialty decisions. This is the case since the higher the fee level, the less work for the same amount of income, or, alternatively, the more income for the same amount of work.

However, physicians generally earn high incomes.[12] Moreover, it is commonly thought that physicians possess the ability to expand, at least to some extent, the demand for their services, which in turn contributes to their ability to maintain and increase their income levels. Hence, the actual or potential income level of the physician may be high enough to dim the attractiveness of higher fees.

Equally important, both locational and specialty choices are among the most important decisions that the physician makes in his or her life. Consequently, physicians will weigh a number of factors in their decisions. The case mix and patient mix handled by the physician, the service mix delivered by the physician, the physician's professional prestige, and the physician's feeling of autonomy will all be heavily influenced by the specialty that the physician chooses. These factors will also be affected by the location decisions, although probably not to the same degree. Another crucial factor in geographic decisions is the characteristics of the locations that the physician is considering.

A crucial point in regard to the latter is that this study is limited to metropolitan areas with the characteristics previously described. This means that the range of variation in locational characteristics is reduced, since very large metropolitan areas and rural areas are excluded from the sample. However, there is also considerable variation in locational characteristics across the SMSAs in this study. For example, all regions of the country are included; there is a wide range in population, from approximately 50,000 to about 1.5 million; and there are wide climatic differences among the SMSAs.

Nonetheless, locational characteristics would be expected to have less impact in terms of the sample of 91 SMSAs compared to the country as a whole.

It seems likely that the influence of physician fee levels will be highest for those physicians who shift location and/or specialty in mid-career. These physicians are familiar with the business and administrative aspects of medical practice. Consequently, it is probable that they will ascertain and consider differences in physician fee levels across metropolitan areas. Differential physician fee levels are also likely to have an impact on those physicians entering practice, especially since the income offers that such physicians receive from established practices will presumably vary directly with the fee levels in the community.

Physician fee levels are likely to be least important for those physicians leaving practice, since such physicians will presumably have adjusted their expectations to prevailing community fee levels. This would not be the case if community fee levels were sharply reduced or if a freeze were imposed on fee levels for a substantial period of time. However, during the period of this study neither of these situations occurred.[13]

Physician fees would be expected to be an important consideration for both USMGs and FMGs. However, they may be relatively more important for FMGs because other factors, such as medical education, may be less important. This supposition is generally supported by the findings of Hadley (1975), which will be discussed shortly. Nonetheless, since those factors most heavily influencing the distribution of FMGs have not yet been fully identified, the hypothesis that fees may affect FMGs more than USMGs must be regarded as tentative at this point. However, it does seem likely that physician fee levels should have at least some significant influence on both FMGs and USMGs.

It would have been preferable in this study to have examined the impact not only of gross physician fee levels, but also of net physician fee levels, adjusted for the probable business expenses of physicians. However, in the first phase of this study, it was not possible to obtain health personnel wage indices across the sample of SMSAs. Nonetheless, Steinwald and Sloan (1974) demonstrated that the wages of ancillary personnel are the best predictors of physician fees, both in terms of significance levels and elasticities. This may mean that adjusting physician fees for ancillary personnel wages does not significantly affect the impact of fees on physician distribution.

There is relatively little direct evidence from previous studies about the impact of physician fee levels on the distribution of physicians. However, Newhouse and Phelps (1974) found a positive relationship between the concentration of physicians and the price of clinic visits times the mean coinsurance rate for such visits. In considering the possible associations between physician concentrations and physician fee levels, it is important to examine the following three relationships:

(1) the relationship between physician concentration and physician fee levels at a point in time;
(2) the relationship between physician fee levels at a point in time and the change in the concentration of physicians over a subsequent period of time; and
(3) the relationship between the concentration of physicians at a point in time and the change in physician fee levels over a subsequent period of time.

It seems likely that relationship (1) is highly positive, i.e., there is a strong positive association between the concentration of physicians and physician fee levels. However, this positive association in and of itself does not establish either relationship (2) or (3). It is possible that the concentrations of physicians and physician fee levels are influenced by similar factors and that they do not directly affect each other significantly. In this case relationship (1) would be positive and relationships (2) and (3) would be insignificant.

However, for the reasons indicated before, it also seems likely that physician fee levels at a point in time will positively influence the change in the concentration of physicians during a subsequent period of time. Consequently, we are hypothesizing that relationship (2) is positive.

Further, there appear to be sound reasons for expecting that relationship (3) is also positive.[14] The primary argument in this regard is that a relatively high concentration of physicians may induce an upward movement in physician fees to compensate for possible decreases in the volume of patients per physician in such areas.[15] As indicated earlier, physician fee data were obtained for only one year in the first phase of this project. Therefore, it is not possible to test relationship (3) in this project.

Surgeons generally earn higher incomes than do primary care physicians.[16] Consequently, surgeons may be less influenced by prevailing physician fee levels than are primary care physicians. However, surgical fees are more visible and often more clear-cut than medical fees. Thus, surgeons may be more aware of differences in fee levels and may be more likely to respond to those differences. It would appear that these two factors would tend to balance each other. Thus it seems likely that physician fee levels will have similar effects on surgeon and primary care physician distribution. In summary, the following hypotheses will be tested:

(1) Physician fee levels will be strongly positively associated with the concentration of physicians.

(2) Physician fee levels will have a substantial positive influence on the change in the concentration of physicians between 1968 and 1973.

(3) Hypotheses (1) and (2) will pertain to office-based physicians, surgeons, and primary care physicians.

MEDICAL EDUCATION

There appears to be a solid conceptual rationale underlying the contention that the location of medical education programs will have a positive influence on the concentration of physicians. This rationale has two primary arguments. First, physicians tend to stay in places where they have been educated. There are two main reasons why this is probably the case: (a) the physician may have selected the training program on the basis of its location, intending to continue in that location after the completion of the training program, and (b) as a physician participates in an educational program, he or she becomes familiar not only with the characteristics of the metropolitan area itself, but also with the intricacies of medical practice in the area. Therefore, it is relatively easy to leave the training program and set up practice in the same area.

The second argument is that physicians tend to choose SMSAs in which a medical center is located, even if they were not educated at the medical centers in question. Some physicians may wish to participate directly in the teaching programs of the medical center. Others may wish to utilize its specialized facilities. Still others may wish to practice in an SMSA in which a medical center is located because intellectual stimulation is thought to be greater in such communities. Thus, practice in SMSAs in which a medical center is located enters a physician's thinking partly because it may influence net income and workload, but primarily because it may affect case mix and patient mix, physician service mix, the quality of physician service (as viewed by the physician), and especially the physician's professional prestige in the physician community, including that community of similar physicians that extends beyond the particular SMSA.

It is likely that the attraction of a medical center is greater for a specialist than for a family or a general practitioner. Specialists are, on the whole, more likely to be sought by medical schools than are family or general practitioners. Specialists will be drawn to the array of facilities in a medical center. The opportunities for family or general practitioners, on the other hand, may be higher in communities in which a medical center is not located because of the emphasis on specialization in many SMSAs with a medical center. Hence we would expect that the impact of the location of medical centers would be greater on surgical specialists than on primary care physicians. However, this hypothesis is attenuated by the fact that, while approximately 50% of primary care physicians are general or family practitioners, the remaining 50% are medical specialists, who may be highly attracted to SMSAs containing medical education programs.

The location of medical education programs may also be an attraction to FMGs. In many instances FMGs come to the United States to take postgraduate medical training, and they may elect to set up practice in the area in which they have trained. However, the strong tendency for physicians who have received undergraduate and postgraduate medical education in the same state to remain in that state is not necessarily present for FMGs.

Equally important, it is likely that there is a tendency for FMGs to fill in the "cracks" in the U.S. medical system (although not necessarily the cracks in rural areas). Further, medical centers, including hospitals only peripherally involved with such centers, may not fully recognize the credentials of FMGs. This is not to say that medical centers do not formally recognize FMG credentials, but that FMGs are likely to be accorded less prestige within the medical center environment than USMGs. Moreover, it seems probable that this tendency will spill over into the community surrounding the medical center.

For all these reasons it seems likely that SMSAs containing medical education programs will be more attractive to USMGs than to FMGs. Given the sharp increase in the entry of FMGs into this country, it thus seems probable that the impact of medical education will be much smaller on the change in the concentration of physicians between 1968 and 1973 than upon the concentrations of physicians in the four study years, since the latter include not only the latest cohort of net physician change but also earlier cohorts across the past 30 to 40 years.

Previous studies are generally supportive of these hypotheses. In a study using unpublished medical economics data from the mid-sixties, Yett and Sloan (1974) found that residency in a state had the greatest impact upon the retention of the physician within the same state. They also found a strong influence of medical school location and internship location on the eventual practice location of physicians.

Scheffler (1971) found a very strong positive correlation between the concentration of medical specialists and the concentration of internships and residencies, and also between the concentration of surgical specialists and the concentration of internships and residencies. Further, Reskin and Campbell (1973) found that the concentration of interns and residents was positively related to the number of physicians, after the effects of population were taken into account.

Balinsky (1974) found that a high percentage of physicians graduating from the University of Buffalo Medical School tended to practice in those areas in which they had taken postgraduate training. Similarly, Martin et al. (1968) demonstrated a positive relationship between graduate training in Kansas and eventual practice location in that state.

Cooper et al. (1975) found that primary care physicians indicated that opportunity for regular contact with a medical school or medical center was an important influence on their choice of practice locations. However, these authors also determined that medical school or medical center location was not the most important practice location consideration indicated by primary care physicians.

In addition, Weiskotten et al. (1960) examined physician location choices over a 35-year period from 1915 to 1950, utilizing 2-by-2 tables for analysis. They found that the most important factors in practice location in descending order were place of residency, prior residence, place of internship training, and place of medical school education.

Hadley (1975) has undertaken perhaps the most comprehensive study of the impact of medical education on physical distribution, since he considers those factors which tend to be most important for physicians in different education cohorts. He found that physicians who practice in the same state where they undertook their undergraduate and graduate education tend overwhelmingly to have been residents of the same state prior to entry into medical school. He also ascertained that neither the pecuniary nor non-pecuniary characteristics of the state had an important influence on physician choice in these instances. Thus, Hadley contends that physicians in this group choose the state of their medical education because of personal or psychic ties to their home area.

In contrast, physicians who locate in areas in which they took only their graduate training prefer states with relatively high non-pecuniary returns and opportunity for professional contact. Hadley found that such physicians tend to migrate from inland states to coastal states, with the rate of net population migration and the existing concentration of specialists per capita influencing location choices. In further contrast, physicians who locate in a state in which they took no medical education seem to behave in more traditional economic ways—they appear to be attracted by higher expected monetary returns. However, these physicians were also attracted to areas where the physician specialist concentration and population growth rate were higher.

Thus, given Hadley's results, in a cross-sectional and time-series study such as this one, which examines the concentration of all office-based physicians (or the concentrations of surgeons and primary care physicians), one would expect that medical education would be a prime determinant of physician distribution. Further, Hadley's findings also support our previous hypothesis that medical education will have a smaller influence on FMGs than upon USMGs.

In light of Hadley's work, one would also anticipate that physician fee levels would have an important influence on physician distribution, as we have previously hypothesized. Further, it is likely that region will have a major impact on physician distribution, as we will discuss soon. In summary, the following hypotheses will be tested:

(4) The extent of medical education programs will have a positive impact on the concentration of physicians.

(5) The extent of medical education programs will have a substantially smaller and perhaps insignificant influence on the change in the concentration of physicians between 1968 and 1973. However, if its impact is significant, the direction of its influence should remain positive.

(6) Hypotheses (4) and (5) may be somewhat stronger for surgical specialists than for primary care physicians, although this difference may be insignificant.

WEST REGION

As emphasized earlier, the characteristics of the location which the physician is considering is a crucial argument in the physician objective function, especially in location decisions. Some of these characteristics are socioeconomic and demographic.[17] However, other characteristics are better captured by the region and climate variables. Some of the latter characteristics are relatively straightforward, such as climate and recreational opportunities. Others are harder to define, but are nonetheless important, such as life-style differences.

Further, there are also important regional differences in the style of medical practice. Such differences may affect the location decisions of physicians through such factors as physician case mix and patient mix, physician service mix, the quality of physicians (as viewed by him or her), and the physician's feeling of autonomy.

Therefore, it seems likely that the impact of region and climate will be substantial on the geographic and specialty distribution of physicians. Moreover, region and climate are likely to affect all three cohorts of physicians discussed earlier. Regional and climatic considerations may be important to the new physician entering practice; they may also be the controlling determinant in the decisions of some physicians to shift practice location in mid-career. Regional and climatic considerations may induce physicians to continue to practice to a later age in some areas than in others. In addition, region may have a delayed impact on this third cohort. For example, if a particular region were popular among physicians 35 years ago, then that region may be negatively associated with the change in the concentration of physicians at the present time, as those physicians who entered practice three to four decades ago now leave the medical field.

There are substantial differences in the regional patterns of location among USMGs and FMGs (U.S. DHEW, 1974). For example, in 1970, 10% of California physicians were FMGs. Considering other Western states, 7% of physicians in Colorado were FMGs, as were 11% of physicians in Arizona and 11% of physicians in Washington. In contrast, in the Midwestern states of Illinois, Ohio, and Michigan, the percentages of physicians who were FMGs were, respectively, 29%, 25%, and 25%. The penetration of FMGs in the Eastern states also tends to be high. New York, where 38% of physicians are FMGs, is highest of all the states on this dimension. However, it is likely that FMGs are disproportionately concentrated in New York City, as opposed to the five upstate New York SMSAs in the sample.[18]

At a minimum, it seems clear that the West/Midwest axis along which USMGs tend to be distributed (i.e., relatively high concentrations of physicians in the West and relatively low concentrations in the Midwest) does not hold for FMGs. Therefore, given the increase in the inflow of FMGs during the last five years, it seems likely that the impact of the West region on the changes in the concentrations of physicians during the study period will be much smaller than the influence of the West region on the concentrations of physicians during the study period. In fact, the effect of the West region may be insignificant.

Relatively few studies have explicitly examined the impact of region and climate on the distribution of physicians. Yett and Sloan (1974) found that a cold climate (measured by the daily difference between mean temperature and 65° F.) was negatively associated with the concentration of physicians. De Vise (1973) pointed out the migration of physicians from the Midwest, especially Illinois, to states on both coasts. Fahs et al. (1968) found somewhat similar results involving a different set of states. In summary, the following hypotheses will be tested:

(7) The West will have a strong positive influence on the concentration of physicians.

(8) The West will have a much smaller and perhaps insignificant effect on the change in the concentration of physicians.

(9) Hypotheses (7) and (8) will hold for office-based physicians, surgical specialists, and primary care physicians.

FINDINGS

CORRELATION FINDINGS

As shown in Table 2, all three independent variables are significantly positively associated with the concentrations of the

TABLE 2
CORRELATION FINDINGS[a]

	Surgical Fees	Interns and Residents	West
Office-based physicians	.65 to .69	.41 to .47	.54 to .58
Surgeons	.47 to .54	.35 to .40	.38 to .42
Primary care physicians	.66 to .69	.28 to .37	.52 to .59
Δ Office-based physicians	.40	.07	.23
Δ Surgeons	.35	.01	.13
Δ Primary care physicians	.25	.06	.08

a. The correlations indicated for the physician concentration measures encompass the range of correlations across the four study years.

three types of physicians, at greater than the .01 level. Further, this finding holds for all four study years.

The correlation between the surgical fee index and the physician concentration measures is higher than the correlations involving the other two independent variables. Although there is some overlap, the correlation between the West and the physician concentration measures tends to be stronger than the correlation between the medical education measure and the physician concentration variables.

While the extent of association is lower, the surgical fee index is also highly significantly correlated with all three physician change variables. In marked contrast, the medical education measure is significantly correlated with none of the physician change variables. The West region is significantly positively correlated with the office-based physician change variable, but not with the other two physician change variables.

PROFILE ANALYSES

As shown in Table 3, the profile analyses confirm the correlation findings involving the independent variables and the physician concentration measures. All three of the independent variables are highly significantly positively related to the three physician concentration measures.

The surgical fee index is highly significantly positively associated with the change in the concentration of office-based physicians, and significantly positively associated with the change in the concentration of surgeons. However, it is insignificantly associated with the change in the concentration of primary care physicians.

Corresponding to the correlation findings, the concentration of

TABLE 3
PROFILE ANALYSES[a]

	Surgical Fees	Interns and Residents	West
Office-based physicians	.00/.00	.00/.00	.00/.00
Surgeons	.00/.00	.00/.00	.00/.00
Primary care physicians	.00/.00	.00/.00	.00/.00
Δ Office-based physicians	.00/.01	.93/.64	.10/.13
Δ Surgeons	.02/.01	.80/.36	.12/.22
Δ Primary care physicians	.51/.80	.26/.69	.65/.81

NOTE: The profile variables are located across the top of this table. The filter variables are shown along the left side of the table. For each filter variable, the sample of 80 SMSAs was divided in terms of the highest one-third of SMSAs and the lowest one-third of SMSAs in regard to each specific filter variable.

a. The first column in each entry indicates the p-value for the normal deviate test, which is a parametric profile procedure. The second column indicates the p-value for the Wilcoxon test, which is a non-parametric profile technique. The p-values are interpreted as indicating the significance level at which a difference exists between the two groups included in the profile. Thus, a p-value of .05 indicates that there is a difference, in the profile variable between the two groups included in the analysis, that is significant at the .05 level.

interns and residents is not significantly related to any of the physician change variables. The same result pertains to the West region, although the p-values on the normal deviate and Wilcoxon tests are, respectively, .10 and .13 for the relationship between the West and the office-based physician change variable. However, the level of significance in this relationship is not as high as in the correlation analyses.

REGRESSION RESULTS

Office-Based Physicians. All three independent variables are highly significant predictors of the concentration of office-based physicians in 1973 and 1970 (see Appendix A). The t-values for the three independent variables are highly significant at substantially beyond the .01 level. The R^2 s in these two equations approximate .70, i.e., these three independent variables explain 70% of the variation in the concentration of office-based physicians.

In these two equations the elasticity of the surgical fee index approximates +.70, while the elasticities of the other two variables are in the range of +.06. Moreover, it would appear at least as easy in a real world setting to increase or decrease the fee index by a given percentage amount as to change either of the other variables to the same extent.

In these two equations the marginal R^2 of the West region is the lowest of the three variables. The marginal R^2s of the other two variables are approximately the same in the 1970 equation, while the marginal R^2 of the surgical fee index is greater in the 1973 equation.[19]

In marked contrast, only the surgical fee index is a significant predictor of the change in the concentration of office-based physicians. Its t-value is significant at greater than the .01 level, and its sign is positive, as expected. Its elasticity is +3.6, i.e., substantially higher than its elasticity in the physician concentration equations. However, this is expected given the much lower mean of the physician change variable as compared to the physician concentration measure.

As expected, since two of the three independent variables are not significant predictors in this equation, the R^2 is much lower, approximating .17. This lower R^2 is also due to the fact that the surgical fee index is not as powerful a predictor of the physician change variable as it is of the physician concentration measure.

Surgeons. As with office-based physicians, all three variables are significant positive predictors of the concentrations of surgeons in 1973 and 1970 (see Appendix B). The t-values of the surgical fee index and the intern and resident measure are highly significant at greater than the .01 level. In slight contrast, the t-value of the West region is significant at the .01 level in the 1970 equation but only at the .05 level in the 1973 equation.

Reflecting the generally lower t-values, the predictive power of these three variables in terms of the concentration of surgeons is lower than it is in terms of the concentration of office-based physicians. The R^2 in the 1973 and 1970 equations are, respectively, .44 and .40.

The elasticities of the three variables are similar to those in the office-based physician concentration equations. The surgical fee index has an elasticity of +.64 in the 1973 equation and +.48 in the 1970 equation. The elasticity of the intern and resident measure approximates +.07, while the elasticity of the West region approximates +.04.

The marginal R^2 of the surgical fee index in the 1973 equation is substantially higher than its marginal R^2 in the 1970 equation. The marginal R^2 of the intern and resident measure is approximately the same in the two equations, being therefore lower than that of the surgical fee index in the 1973 equation and higher in the 1970 equation. The marginal R^2 of the West region is the lowest in both equations.

The surgical fee index is the only significant predictor of the surgeon change variable. Its elasticity is +2.6, somewhat lower than its elasticity in terms of the office-based physician change variable, but in the same general range. The same applies to the R^2 of this equation, which is .12, as compared to .17 for the office-based physician change equation.

Primary Care Physicians. The findings regarding primary care physicians are similar to those involving office-based physicians and surgeons (see Appendix C). As before, all three independent variables are significant predictors of primary care physician concentration in 1973 and 1970. Further, as with office-based physicians, the t-values of the three independent variables are significant in each instance at substantially beyond the .01 level.

The predictive power of these two equations stands intermediate between that of the office-based physician equations and that of the surgeon equations. The R^2 in the 1973 equation is .63 and in the 1970 equation .60.

The elasticities of the three independent variables are also similar to their previous elasticities. The elasticity of the surgical fee index is +.80 in the 1973 equation and +.68 in the 1970 equation. The elasticities of both the intern and resident measure and the West region approximate +.05.

Although the marginal R^2 of the surgical fee index is lower in the 1970 equation than in the 1973 equation, this variable still has the highest marginal R^2 in both equations. In both equations the marginal R^2 of the West region is higher than the marginal R^2 of the intern and resident measure.

As with the office-based physician and surgeon change equations, the only significant predictor of the change in the concentration of primary care physicians is the surgical fee index. Its t-value is 2.13, which is significant at the .05 level, but not at the .01 level. However, its elasticity is higher in this equation than in the other two physician change equations, being +6.5.

Summary. All three independent variables are strong positive predictors of the concentrations of all three types of physicians. As a group they predict most strongly office-based physician concentration, next most strongly primary care physician concentration, and least strongly surgeon concentration. However, even in the case of the latter, the R^2 is substantial.

In the physician concentration equations the surgical fee index tends to be the strongest independent variable, both in terms of elasticity and marginal R^2. The elasticities of the other two independent variables are highly similar. In addition, the marginal R^2 of the medical education measure tends to be somewhat higher than the marginal R^2 of the West region, especially for surgeons.

In contrast, the surgical fee index is the only significant predictor of the physician change variables. Its predictive power is less in terms of the physician change variables than in terms of the physician concentration measures. Nonetheless, the elasticities of the surgical fee index in the physician change equations are sufficiently high to give that index real-world importance as well as statistical significance.

RELATIONSHIPS WITH OTHER INDEPENDENT VARIABLES

There is a strong bivariate positive association between the West region and the surgical fee index. The ordinary correlation coefficient between these two variables is .45. This finding is confirmed by the results of profile analyses, which also indicate a highly significant positive association between the surgical fee index and the West region. In contrast, the intern and resident measure is not significantly associated with either the surgical fee index or the West region on either correlation or profile analyses.

These same points are indicated by the covariance ratios of these three variables in the regression equations. The covariance ratio of the medical education measure is very low, indicating essentially no relationship with the other two independent variables. The covariance ratio of the other two measures approximates .21, which indicates a significant positive association between these two variables; however, this association is not so high that it negates the regression results.

The determinant in the regression equations approximates .78. The p-value for the chi-square test for overall collinearity is 0.0 for all of the equations. Therefore, the regression results are relatively free of collinearity effects.

STRATIFIED ANALYSES

Nonetheless, because of the positive association between the West region and the surgical fee index, it appeared desirable to conduct regression analyses using a stratified sample excluding the SMSAs in the West region. The purpose of this analysis was to further separate the effects of the surgical fee index from the West region by determining the influence of that index in the non-Western part of the United States. However, since this was a supplementary analysis, only one set of the stratified analyses is contained in this chapter–those for office-based physicians (see Appendix D).

Both the surgical fee index and the intern and resident measure are strong positive predictors of the concentrations of office-based physicians in 1973 and 1970 in the stratified sample. The t-values for both of these variables are significant at substantially beyond the .01 level. Further, the R^2 in these two equations (which contained only the two variables) are high, being .64 and .60, respectively, in the 1973 and 1970 equations.

The elasticities of the two variables in the stratified analyses are similar to their elasticities in the analyses using the entire sample. The elasticity of the fee index is +.84 in 1973 and +.62 in 1970. The elasticity of the intern and resident measure is the same in the two equations, being +.07.

The surgical fee index is a strong positive predictor of the change in the concentration of office-based physicians in the stratified equation. As in the equation using the entire sample, the intern and resident measure is an insignificant predictor. The t-value of the surgical fee index in this equation is significant beyond the .01 level. In addition, the R^2 of this equation is greater than the R^2 of the equation using the entire sample, being .21.

Thus, the findings of this stratified analysis confirm the results using the entire sample, in that the fee index and the intern and resident measure are strong positive predictors of the concentration of office-based physicians, but only the fee index is a significant predictor of the office-based physician change variable.[20]

SUMMARY AND POLICY IMPLICATIONS

The findings of this study largely support the hypotheses tested. There are two relatively minor exceptions to this point. The initial hypotheses indicated that both the intern and resident measure and the West region would have a substantially smaller and possibly insignificant impact on the physician change measures. The findings showed that these relationships are in fact insignificant. Consequently, these two hypotheses must be modified to take into account this result.

The hypotheses (developed initially in the section "Conceptual Framework") are restated below in their modified form, taking into account the findings of this project. Thus, they represent a summarization of the results of these analyses.

(1) Physician fee levels are strongly positively related to the concentration of physicians.

(2) Physician fee levels have a substantial positive influence on the change in the concentration of physicians between 1968 and 1973.

(3) The extent of medical education programs has a strong positive impact on the concentration of physicians.

(4) The extent of medical education programs has an insignificant influence on the change in the concentration of physicians.

(5) The West region has a strong positive effect on the concentration of physicians.

(6) The West region is insignificantly related to the change in the concentration of physicians.

(7) All these hypotheses apply to office-based physicians, surgical specialists, and primary care physicians.

These findings have a number of important policy implications. The finding that all three of these variables are strong positive predictors of physician concentration indicates that a network of generic factors is important in determining physician distribution. The influence of fees indicates that physicians tend to respond to financial incentives. The impact of the medical education measure means that physicians also consider the professional environment in which they will be practicing. The effect of the West region indicates that the characteristics of the overall environment are also highly important in physician location decisions.

These points are underscored by the fact that the overall physician analyses in this project indicate that these three variables are the strongest predictors of physician concentration. This is especially significant since a number of other variables were included in the overall physician analyses in order to test additional hypotheses. Among such variables are public insurance measures, private insurance indices, concentration of hospital beds, changes in the concentration of hospital beds, a physician licensure stringency index, a medical malpractice insurance index, hospital interaction measures, the HMO penetration rate, the income profile of the SMSA, and other socioeconomic and demographic variables.

Thus, it appears that the effects of financial, educational, and regional factors are predominant in influencing physician distribution, although the impact of medical education and the West region seem to be waning, due to the influence of FMGs, as will be discussed shortly.

Another implication derives from the strong positive influence of physician fee levels, not only on the concentration of physicians, but also on the change in the concentration of physicians. This finding supports, but does not establish, the contention that physician fee levels can be effective and efficient tools for producing desired alterations in physician distribution. This point is relevant not only to possible future extensions in public or publicly mandated health insurance programs but also to the utilization of existing insurance programs as regulatory tools. This is the case because it is relatively easy in broad insurance programs to design physician fee structures in such a way that they represent positive or negative incentives for certain types of physician geographic and/or specialty distributions.

However, in this regard it must be recalled that this project involves only the SMSAs with the characteristics previously described. These SMSAs can be looked upon as representing the "middle third" of this country, i.e., that one-third of the U.S. population that lives neither in the very largest SMSAs nor in towns or rural areas. Thus, this project indicates that physician fee levels appear to have an important impact on physician distribution across these SMSAs. The findings of this project do not indicate that physician fee levels would be effective tools for altering the rural/urban distribution of physicians. It may be that they would be effective tools for this purpose, but this project provides no information on this point.

The strong positive impact of medical education programs also has policy implications. It indicates that the location of medical education programs could perhaps be used as a device for altering physician distribution. Moreover, the finding that the extent of medical education programs does not significantly affect the change in the concentration of physicians between 1968 and 1973 does not necessarily negate the usefulness of medical education programs in altering physician distribution. This is the case since the presumed reason for this lack of impact is the influx of foreign medical graduates, who may be less responsive to the location of medical education programs.

However, in evaluating the usefulness of medical education programs as a policy tool for influencing physician distribution, three points must be kept in mind. The first is that the findings of this study show that the elasticity in this relationship is relatively low. Therefore, if this elasticity is valid (at a minimum, it was highly consistent not only in the analyses discussed in this chapter but also in the other physician analyses of this project), this means that the extent of medical education program must be changed markedly in order to influence substantially the distribution of physicians. Nonetheless, it would be possible for selected individual SMSAs–but presumably not for all SMSAs, without overexpanding medical education programs–to increase substantially the size of medical education programs.[21]

The second point is that, during the last quarter century, residencies and internships have tended to be in oversupply. Thus, simply opening a residency or internship program in an SMSA may not attract physicians to that area, since the residency or internship program may go unfilled.

The third point is analogous to a point made in regard to physician fee levels. This study shows that medical education programs affect physician distribution across these SMSAs. It does not demonstrate that medical education programs can be used to alter the rural/urban distribution of physicians.

Another implication arises from the effects of the West region in these analyses. The impact of this region is strong, but it is not so powerful as to override the influence of other variables which may be easier to modify from a policy perspective.[22] This is perhaps best illustrated by the elasticity of the West region, which approximates +.05. This means that if the penetration rate of Western SMSAs in this sample were to increase by 10%, the regression analysis would predict that the concentrations of the three types of physicians would increase by approximately .5%.

This point is further accentuated by the findings of the entire set of physician analyses. These results show that the income profiles of the SMSAs and other commonly used socioeconomic and demographic measures are not important predictors of physician concentration. The results also show that the Midwest region is a strong negative predictor of physician distribution, especially the concentration of surgeons, and that mean January temperature is a strong positive predictor of the concentration of physicians. However, neither of these regional variables is as powerful as the West region in terms of the concentrations of physicians. Mean January temperature does, however, have an important influence on the physician change variables, unlike the West.

Thus, there is no other situational variable used in this study which had as strong an impact as did the West region on the concentration of physicians. Consequently, in terms of physician distribution across the SMSAs of this project, it appears that situational variables do not fully override the impact of variables which can be at least potentially changed by policy initiatives. Therefore, it would seem that public and private policy changes could substantially affect the distribution of physicians across these SMSAs.

Another important implication arises by inference, rather than as a result of the direct findings of this project. There is no question that there has been a sharp increase in the influx of FMGs into this country. Table 1 makes this point clear. The findings of this study indicate that the factors influencing the location decisions of FMGs are probably substantially different from the factors affecting the location decisions of USMGs. The evidence that most strongly supports this hypothesis is the differential impact of medical education programs on the physician change variables as compared to the physician concentration measures. Further support arises from the analogous differential effects of the West region.

Consequently, even more so than in this study, in future studies of physician distribution it will be necessary to consider carefully the rate of influx of FMGs and to identify those factors that appear to influence the location decisions of these physicians. The entire set of physician analyses conducted in this project has attempted to do this in part. For example, the findings of these analyses suggest that the stringency of physician licensure and the relative openness of the hospital medical staff arrangement have a stronger influence on FMGs than on USMGs.[23]

In the current phase of this project no data were included which directly specify the concentration of FMGs. In the next phase, such measures will be included, and it will thus be possible to investigate further the impact of FMGs on physician distribution patterns in the United States.

APPENDICES

The mean and the standard deviation are given for the dependent variable in each case. The other six statistics given at the top of each regression equation apply to the equation as a whole. The five statistics given in the table at the bottom of each regression equation represent values for each independent variable indicated.

APPENDIX A

CONCENTRATION OF OFFICE-BASED PHYSICIANS, 1973

Mean:	94.602	Standard Deviation:	20.825	R^2:	0.701	F-Value:	59.427
Standard Error:	11.608	Constant:	14.527	Determinant:	0.792	P-Value:	0.000

Independent Variables	Regression Coefficient	Elasticity	T-Value	Marginal R^2	Covariance Ratio
Surgical fee index	0.1793	0.732	7.36	0.213	0.208
Interns and residents/100,000	0.3921	0.068	6.03	0.143	0.004
West region	14.6785	0.047	4.62	0.084	0.205

CONCENTRATION OF OFFICE-BASED PHYSICIANS, 1970

Mean:	92.215	Standard Deviation:	20.017	R^2:	0.681	F-Value:	53.960
Standard Error:	11.536	Constant:	25.593	Determinant:	0.784	P-Value:	0.000

Independent Variables	Regression Coefficient	Elasticity	T-Value	Marginal R^2	Covariance Ratio
Surgical fee index	0.1438	0.602	5.91	0.147	0.215
Interns and residents/100,000	0.4549	0.068	6.12	0.158	0.014
West region	16.0900	0.052	5.10	0.109	0.206

CHANGE IN THE CONCENTRATION OF OFFICE-BASED PHYSICIANS, 1968-1973

Mean:	4.980	Standard Deviation:	7.422	R^2:	0.165	F-Value:	5.016
Standard Error:	6.914	Constant:	−13.163	Determinant:	0.784	P-Value:	0.000

Independent Variables	Regression Coefficient	Elasticity	T-Value	Marginal R^2	Covariance Ratio
Surgical fee index	0.0459	3.559	3.15	0.109	0.215
Interns and residents/100,000	0.0091	0.025	0.20	0.000	0.014
West region	0.9798	0.059	0.52	0.003	0.206

APPENDIX B

CONCENTRATION OF SURGEONS, 1973

Mean:	32.410	Standard Deviation:	7.585	R^2:	0.440	F-Value:	19.898
Standard Error:	5.787	Constant:	8.519	Determinant:	0.792	P-Value:	0.000

Independent Variables	Regression Coefficient	Elasticity	T-Value	Marginal R^2	Covariance Ratio
Surgical fee index	0.0539	0.642	4.43	0.145	0.208
Interns and residents/100,000	0.1256	0.064	3.87	0.111	0.004
West region	3.4432	0.032	2.17	0.035	0.205

CONCENTRATION OF SURGEONS, 1970

Mean:	31.567	Standard Deviation:	7.584	R^2:	0.396	F-Value:	16.637
Standard Error:	6.008	Constant:	13.053	Determinant:	0.784	P-Value'	0.000

Independent Variables	Regression Coefficient	Elasticity	T-Value	Marginal R^2	Covariance Ratio
Surgical fee index	0.0392	0.479	3.09	0.076	0.215
Interns and residents/100,000	0.1534	0.067	3.97	0.125	0.014
West region	4.2751	0.041	2.60	0.054	0.206

CHANGE IN THE CONCENTRATION OF SURGEONS, 1968-1973

Mean:	2.270	Standard Deviation:	2.546	R^2:	0.122	F-Value:	3.510
Standard Error:	2.433	Constant:	–3.551	Determinant:	0.784	P-Value:	0.000

Independent Variables	Regression Coefficient	Elasticity	T-Value	Marginal R^2	Covariance Ratio
Surgical fee index	0.0154	2.615	3.00	0.104	0.215
Interns and residents/100,000	–0.0049	–0.029	0.31	0.001	0.014
West region	–0.1634	–0.022	0.25	0.001	0.206

APPENDIX C

CONCENTRATION OF PRIMARY CARE PHYSICIANS, 1973

Mean:	42.225	Standard Deviation:	9.870	R^2:	0.632	F-Value:	43.541
Standard Error:	6.103	Constant:	4.108	Determinant:	0.792	P-Value:	0.000

Independent Variables	Regression Coefficient	Elasticity	T-Value	Marginal R^2	Covariance Ratio
Surgical fee index	0.0879	0.804	6.86	0.228	0.208
Interns and residents/100,000	0.1324	0.052	3.87	0.073	0.004
West region	6.6935	0.048	4.01	0.078	0.205

CONCENTRATION OF PRIMARY CARE PHYSICIANS, 1970

Mean:	42.664	Standard Deviation:	9.864	R^2:	0.601	F-Value:	38.221
Standard Error:	6.350	Constant:	9.300	Determinant:	0.784	P-Value:	0.000

Independent Variables	Regression Coefficient	Elasticity	T-Value	Marginal R^2	Covariance Ratio
Surgical fee index	0.0753	0.681	5.62	0.166	0.215
Interns and residents/100,000	0.1374	0.044	3.36	0.059	0.014
West region	8.0362	0.056	4.62	0.112	0.206

CHANGE IN THE CONCENTRATION OF PRIMARY CARE PHYSICIANS, 1968-1973

Mean:	–1.067	Standard Deviation:	4.078	R^2:	0.065	F-Value:	1.772
Standard Error:	4.020	Constant:	–8.025	Determinant:	0.784	P-Value:	0.000

Independent Variables	Regression Coefficient	Elasticity	T-Value	Marginal R^2	Covariance Ratio
Surgical fee index	0.0180	6.527	2.13	0.056	0.215
Interns and residents/100,000	0.0079	0.101	0.30	0.001	0.014
West region	–0.3923	–0.110	0.36	0.002	0.206

APPENDIX D

These stratified samples include only those SMSAs in the group of 80 that are not in the West region. There are 56 such SMSAs.

CONCENTRATION OF OFFICE-BASED PHYSICIANS, 1973 (STRATIFIED SAMPLE)

Mean:	86.889	Standard Deviation:	16.684	R^2:	.641	F-Value:	47.305
Standard Error:	10.184	Constant:	7.743	Determinant:	.973	P-Value:	0.000

Independent Variables	Regression Coefficient	Elasticity	T-Value	Marginal R^2	Covariance Ratio
Surgical fee index	.1969	.838	6.69	.303	.027
Interns and residents/100,000	.3926	.073	5.86	.233	.027

CONCENTRATION OF OFFICE-BASED PHYSICIANS, 1970 (STRATIFIED SAMPLE)

Mean:	84.648	Standard Deviation:	15.298	R^2:	.600	F-Value:	39.802
Standard Error:	9.853	Constant:	26.295	Determinant:	.957	P-Value:	0.000

Independent Variables	Regression Coefficient	Elasticity	T-Value	Marginal R^2	Covariance Ratio
Surgical fee index	.1411	.617	4.91	.182	.043
Interns and residents/100,000	.4624	.073	6.26	.296	.043

CHANGE IN THE CONCENTRATION OF OFFICE-BASED PHYSICIANS, 1968-1973 (STRATIFIED SAMPLE)

Mean:	3.868	Standard Deviation:	7.258	R^2:	.207	F-Value:	6.898
Standard Error:	6.586	Constant:	–21.227	Determinant:	.957	P-Value:	0.000

Independent Variables	Regression Coefficient	Elasticity	T-Value	Marginal R^2	Covariance Ratio
Surgical fee index	.0669	6.397	3.48	.182	.043
Interns and residents/100,000	.0264	.091	.53	.004	.158

NOTES

1. The author is also the principal investigator of that larger project, which has involved a number of Spectrum Research staff members. Those most involved in the physician analyses and the preparation of this chapter, in addition to the author, are David K. DeWolf, Nancy Hoffman, Steven Lazarus, Peter Shaughnessy, Garry Toerber, Jean Bell, Alice Bryant, and Tonie Gatch.

2. The central product of this phase, "Factors Influencing Health Care Resources and Utilization in Metropolitan Areas," is now in the preparation stage and should be forthcoming by early summer of 1976.

The Phase I reports of this project discuss the entire set of findings in the physician analyses, as well as the findings of the other analyses. Readers interested in reviewing this wider set of findings can do so by obtaining these reports from Spectrum Research, Inc., 789 Sherman, Suite 500, Denver, Colorado 80203. As indicated earlier, they will be available in published form by early summer 1976.

3. A list of these 91 SMSAs is contained in the larger reports describing this study.

4. Measure (1) was so defined for several reasons. First, the number of federal physicians was excluded from this physician measure in a manner analogous to the exclusion of federal hospital beds from the hospital measures used elsewhere in this study. Second, since the focus here is on the delivery of health care services, the number of non-patient care physicians is excluded from the overall physician measure. Third, after substantial deliberation, the number of hospital-based physicians was excluded from the overall physician measure since the objective was to concentrate on those physicians involved directly in patient care on essentially a full-time basis.

Included within the surgical specialties of measure (2) are: general surgery, neurosurgery, obstetrics and gynecology, ophthalmology, orthopedic surgery, otolaryngology, plastic surgery, colon and rectal surgery, thoracic surgery, and urology.

5. This concern has recently led the surgical specialty societies to restrict the number of surgical residencies.

For ease of exposition, in subsequent sections of this chapter, office-based, patient care, non-federal physicians will be referred to as office-based physicians; office-based, patient care, non-federal physicians practicing in surgical specialties as surgical specialists or surgeons; and office-based, non-federal physicians practicing in general practice or in the medical specialties as primary care physicians.

6. It should be pointed out that this latter premise is largely untested, but the fact that it may be true further accentuates the desirability of examining carefully the determinants of the concentration of primary care physicians.

7. The source of the data in this paragraph is United States Department of Health, Education and Welfare (1974). The term United States medical graduates (USMGs) customarily includes Canadian medical graduates as well as strictly U.S. medical graduates.

8. The penetration of U.S.-born among FMGs has always been low, but has decreased further in recent years.

9. A higher percentage of FMGs than USMGs enter hospital-based practice, but this sharp increase in the immigration of FMGs applies to both office-based practice and hospital-based practice.

10. Other medical education indices were also used in this study. None was found to have as strong an influence on the concentration of physicians as did the intern and resident measure. Further, the correlation between the intern and resident measure and the presence of a medical school is very high (.85), so that the effects of these two variables are similar.

11. Since the surgical fee index is a composite weighted index, it is relatively

meaningless in absolute value terms. Nonetheless, as for all variables, it is important to review the relationship between the mean and the standard deviation. Hence, the mean and standard deviation of this variable are included for this purpose.

It seemed reasonable for two reasons to include the surgical fee index in the 1970 and 1972 independent variable data sets as well. First, it seems likely that the *relative* magnitude of physician fee levels among the SMSAs will remain reasonably constant across a three-year period. However, it seems less tenable to extend this assumption to five years. Consequently, the 1973 surgical fee index was not included in the 1968 independent variable sets. A second reason supporting the inclusion of the fee index in the 1972 and 1970 data sets is that the Economic Stabilization Program tended to stabilize physician fees from late 1971 to the end of 1973.

12. The average annual income for physicians in 1973 was $41,789 (AMA, 1973).

13. It is true that the Economic Stabilization Program first placed a freeze on physician fees for three months and then regulated fees for a little more than two years. However, the ESP was applied relatively uniformly across the country. This type of physician fee regulation would not be expected to produce important differences across the SMSAs.

14. Since we have postulated that all three of the relationships specified are likely to be positive, and since it has not been observed that physician fee levels and the concentration of physicians are exploding to infinity, it is likely that a fourth hypothesis also holds: while the impact of physician fee levels on the net influx of physicians will be positive, and while the influence of the concentration of physicians on the change in fee levels will also be positive, both of these relationships will have a negative second derivative. This means that rather than physician concentrations and fee levels exploding to infinity, they will move to a stable equilibrium, assuming that sufficient time transpires without other shocks to the system.

15. In their reply to Newhouse (1970), Frech and Ginsburg (1972) discuss, among other things, relationships (2) and (3). They argue that relationship (2) should be positive and that relationship (3) should be negative. However, their article is conceptual in nature and does not present analyses bearing on either contention.

16. In 1970, general practitioners earned an average of $33,859 per year, compared to $50,701 for surgeons (AMA, 1973).

17. With minor exceptions, other analyses of this study indicate that the standard socioeconomic and demographic measures are insignificant predictors of physician distribution.

18. As with the specialty penetration figures, these statistics include interns and residents and hospital-based physicians, as well as office-based physicians. However, while they are not directly applicable to this study, it is likely that they reflect differences in the regional distribution of office-based physicians.

19. Regression equations using the 1972 variables are highly similar to the 1973 and 1970 equations. The 1968 equations are naturally different, since the surgical fee index was not extended backward in time to that year. The same is also true for the surgeon and primary care physician concentration equations.

20. The results of the stratified analyses using the surgeon and primary care physician variables also confirm the earlier regression findings.

21. A concrete illustration of this point is that the average concentration of interns and residents in the 91 SMSAs increased by approximately 50% between 1968 and 1973.

22. Although region would appear totally fixed and independent of policy manipulation, in fact this is not entirely the case. Some regional characteristics, such as style of medical practice, may be transportable to other regions. Other characteristics are essentially immutable, such as mean January temperature.

23. The stringency of physician licensure is significantly related only to the surgeon change variable. As expected, the nature of the hospital medical staff has a significant influence on the surgeon and office-based physician change variables, but not on the primary care physician change measure.

REFERENCES

American Medical Association (1973, 1972, 1971, 1968) Directory of Approved Internships and Residencies. Chicago: American Medical Association.

——— (1973) Profile of Medical Practice, 1973. Chicago: American Medical Association, Center for Health Services Research and Development.

BALINSKY, W. L. (1974) "Distribution of young medical specialists from western New York." Medical Care 12: 437-444.

COOPER, J. K., K. HEALD, M. SAMUELS, and S. COLEMAN (1975) "Rural or urban practice: factors influencing the location decision of primary care physicians." Inquiry 12: 18-25.

DE VISE, P. (1973) "Physician migration from inland to coastal states: antipodal examples of Illinois and California." Journal of Medical Education 48: 141-151.

FAHS, I. J., K. INGALLS, and W. R. MILLER (1968) "Physician migration: a problem in the upper Midwest." Journal of Medical Education 43: 735-740.

FRECH, H. E. and P. GINSBURG (1972) "Physician pricing: monopolistic or competitive: comment." Southern Economic Journal 38: 573-577.

HADLEY, J. (1975) "Models of physicians' specialty and location decisions." Report Number 6, National Center for Health Services Research, Intramural Research Section.

MALONEY, J. V., Jr. (1970) "A report on the role of economic motivation in the performance of medical school faculty." Surgery 68.

MARTIN, E. D., R. E. MOFFAT, R. T. FALTER, and J. D. WALKER (1968) "Where graduates go: the University of Kansas School of Medicine: a study of the profile of 959 graduates and factors which influenced the geographic distribution." Journal of the Kansas Medical Society, pp. 84-89.

NEWHOUSE, J. P. (1970) "A model of physician pricing." Southern Economic Journal 37 (October 19): 174-183.

——— and C. E. PHELPS (1974) "Price and income elasticities for medical care services," pp. 139-193 in M. Perlman (ed.) Economics of Health and Medical Care. New York: John Wiley.

RESKIN, B. and F. L. CAMPBELL (1973) "Physician distribution across metropolitan areas." American Journal of Sociology 79 (July): 981-998.

SCHEFFLER, R. M. (1971) "The relationship between medical education and the statewide per capita distribution of physicians." Journal of Medical Education 46: 995-998.

STEINWALD, B. and F. A. SLOAN (1974) "Determinants of physicians' fees." Journal of Business 47: 493-571.

United States Department of Health, Education and Welfare [DHEW] (1974) Foreign Medical Graduates and Physician Manpower in the United States. Health Resources Administration, Bureau of Health Resources Development, Division of Manpower Intelligence, Office of International Health Manpower Studies, Washington, D.C.: DHEW Publications No. (HRA) 74-30 (February 1974).

WEISKOTTEN, H. G., W. S. WIGGINS, M. GOOCH, and A. TIPNER (1960) "Trends in medical practice: analysis of the distribution and characteristics of medical college graduates, 1915-1950." Journal of Medical Education 35: 1071-1095.

YETT, D. and F. P. SLOAN (1974) "Migration patterns of recent medical school graduates." Inquiry 11: 125-142.

The Authors

ROGER S. AHLBRANDT, Jr., Director of Housing Research and Development for ACTION-Housing, Inc. is an economist and has concentrated on policy development and program evaluation in the field of urban affairs. His background includes experience in the public and private sectors in areas of housing and both corporate and public finance. He has served as a consultant to the U.S. Department of Housing and Urban Development, Commonwealth of Pennsylvania, and various local governments. He has contributed articles to academic and professional journals, has written a monograph, *Municipal Fire Protection Services: Comparison of Alternative Organizational Forms,* and has co-authored a book, *Neighborhood Revitalization: Theory and Practice.* His education includes degrees from Yale University, Harvard Business School (M.B.A.), and University of Washington (Ph.D.).

ROBERT L. BISH is currently Associate Professor and Director of Research for the Institute for Urban Studies at the University of Maryland. He received his Ph.D. in economics from Indiana University in 1968. His publications include *The Public Economy of Metropolitan Areas, Financing Government* (with Harold M. Groves), *Understanding Urban Government: Metropolitan Reform Reconsidered* (with Vincent Ostrom), *Economic Principles and Urban Problems* (with Robert J. Kirk), *Coastal Resource Use: Decisions on Puget Sound* (with Robert Warren, Louis Weschler, James Crutchfield, and Peter Harrison), and *Urban Economics and Policy Analysis* (with Hugh O. Nourse). He has had several articles published.

PHILIP B. COULTER is a senior political scientist in the Center for Population and Urban-Rural Studies at the Research Triangle Institute. He received his Ph.D. in political science from the State University of New York at Albany and has been Associate Professor of Political Science at the University of Massachusetts. His publications include *Social Mobilization and Liberal Democracy, Politics of Metropolitan Areas: Selected Readings,* and *Voting Patterns in Massachusetts: Explorations in Political Ecology,* as well as numerous articles on public policy and decision-making analysis.

LUVERN L. CUNNINGHAM is Executive Director of the San Francisco Public Schools Commission and is currently on two years' leave of absence from the Ohio State University where he is Novice G. Fawcett Professor of Educational Administration and Director of the Institutional Leadership program of the Mershon Center. Cunningham was Dean of the College of Education at the Ohio State University for several years and prior to that, Director of the Midwest Administration Center at the University of Chicago. He has worked actively on the problems of urban schools over the past two decades. From 1973 through June 1975 he directed a large citizens' task force in Detroit. In the 1960s he directed or participated in major studies of school districts in Louisville, Kentucky; Detroit, Michigan; Columbus, Ohio; and Cincinnati, Ohio. His interests focus on problems of governance, management, and organization, and his research and publishing are concentrated in these areas.

JANET FINK is a staff attorney with the New York Legal Aid Society Juvenile Rights Division. She received a B.A. in political science from Bryn Mawr College and a J.D. from Georgetown University Law Center. Dr. Fink has worked as a Junior Fellow at the Metropolitan Applied Research Center, New York, and as a research assistant at the Urban Institute, Washington, D.C., and the Russell Sage Foundation, New York. In 1974, she edited a symposium on juvenile law for the *American Criminal Law Review.*

ROBERT B. HAWKINS, Jr., a recent research fellow at the Hoover Institution, is presently a scholar at the Woodrow Wilson International Center for Scholars in Washington, D.C. He received his Ph.D. in political science from the University of Washington. He has held a number of administrative posts, including those of Director of the California State Office of Economic Opportunity and Chairman of the California Local Government Reform Task Force.

LOIS MacGILLIVRAY, a senior sociologist in the Center for Population and Urban-Rural Studies at the Research Triangle Institute, received her M.A. and Ph.D. degrees in sociology from the University of North Carolina at Chapel Hill. She is Principal Investigator for the NSF/RANN project, "Evaluating the Organization of Service Delivery: Fire," from which the current essay was drawn. Beyond her current work, she was responsible for "Municipal Power Distribution and Municipal Change," a study of the correlates of urban program success.

PATRICK O'DONOGHUE is the President of Spectrum Research, Inc., and of Policy Center, Inc., which are involved in applied health care research. He received an M.D. from Washington University School of Medicine, an M.P.H. from Harvard School of Public Health and a Ph.D. in economics from the University of Washington. Dr. O'Donoghue previously served as a Peace Corps physician in Kuala Lumpur, Malaysia, as an economics instructor at the University of Washington, and as the Associate Director of the Health Services Research Center at InterStudy in Minneapolis. He currently holds an appointment as Adjunct Associate Professor of Medicine at the Boston University School of Medicine. Among his publications in the field of health care are: *Factors Influencing Health Care Research and Utilization in Metropolitan Areas, Evidence About the Effects of Health Care Regulation, Assuring the Quality of Health Care* (with Paul M. Ellwood, Rick J. Carlson, Walter McClure, and Robert Holly), and *A Descriptive Analysis of CHP (b) Agencies* (with Alice Bryant and Peter Shaughnessy).

ELINOR OSTROM is Professor of Political Science and Co-Director of the Workshop in Political Theory and Policy Analysis at Indiana University. Her fields of interest include institutional analysis, urban politics, political methodology, and evaluation research. She is currently Co-Principal Investigator of a RANN-NSF funded study on the effectiveness, efficiency, equity, and responsiveness of organizational arrangements for the provision of police services in metropolitan areas. She is the author of *Urban Policy Analysis: An Institutional Approach* and of journal articles on various aspects of policy analysis.

ROGER B. PARKS is a Research Associate with the Workshop in Political Theory and Policy Analysis at Indiana University. He is currently Co-Principal Investigator with Elinor Ostrom and Gordon P. Whitaker on a National Science Foundation (RANN) funded study of the organization of police service delivery in metropolitan areas across the country. Recent publications include an analysis of the relationship of the experience of victimization and the police response to reported victimization, with citizen perceptions and evaluations of the police services they receive.

DOUGLAS D. ROSE is an Associate Professor of Political Science at Tulane University. His previous work examines aspects of public opinion and public policy, particularly in the American states. He publishes most frequently in the *American Political Science Review,* usually in the form of comments, rejoinders, and replies to communications. He is currently studying representation.

E. S. SAVAS is Professor of Public Systems Management at the Graduate School of Business, Columbia University, and Director of the Center for Government Studies there, as well as Associate Director of the Center for Policy Research. He served as First Deputy City Administrator of New York City and is a consultant to the Committee for Economic Development, the National Science Foundation, and the National Bureau of Standards. He has published widely in the areas of urban management, productivity in local government, urban systems analysis, and urban management information systems. He is Associate Editor of *Management Science.*

MICHAEL P. SMITH is Associate Professor of Political Science at Tulane University. He has taught at Dartmouth College and Boston University. He is co-author of *Politics in America: Studies in Policy Analysis,* editor of *American Politics and Public Policy,* and co-editor of and a contributor to *Political Obligation and Civil Disobedience* and *Organizational Democracy: Participation and Self-Management.* He has contributed extensively to various books and professional journals including the *Journal of Politics, Public Administration Review,* and *Administration and Society.* He is currently working on a book on the impact of urbanization on personality.

WILLIAM EDWARD VICKERY, a senior economist in the Center for Population and Urban-Rural Studies at the Research Triangle Institute, has devoted the majority of his research work to the area of economic development in less developed countries. He has served as a consultant to the governments of Canada and Ghana. He received his M.B.A. from Harvard and his Ph.D. in economic development and economic history from the University of Chicago. A member of the American Economic Association, he has taught at the University of Western Ontario and the University of Ghana. His current work includes research on "Village Development Projects and African Economic Growth" and the measurement of social benefits for farm-to-market roads.

DENNIS R. YOUNG is Associate Professor in the W. Averell Harriman College for Urban and Policy Sciences at the State University of New York at Stony Brook. His research interests focus on the economic organization of public services. He is author of *How Shall We Collect the Garbage: A Study in Economic Organization,* and co-editor and author, with Richard R. Nelson, of *Public Policy for Day Care of Young Children.* He is currently engaged in a study of foster care services delivered by voluntary agencies in New York. Dr. Young received the B.E. and M.S. degrees in electrical engineering from the City College of New York and Stanford University, respectively, and the Ph.D. in engineering-economic systems from Stanford University.

DONALD G. ZAUDERER is Associate Professor in the School of Government and Public Administration at the American University. He is Director of the Public Administration Program and of the Key Executive Program for mid-career professional administrators. He received a Ph.D. in political science at Indiana University and has published in the areas of urban management, voting, and internship education.